SECURITY+

A PRACTITIONER'S STUDY GUIDE

SECURITY+

A PRACTITIONER'S STUDY GUIDE

DAVID LEE EVENDEN
Founder & Owner – StandardUser Cyber Security

LAUREN PROEHL
Information Security Professional

Oversight: Kent Potter
Book Design: Loryn O'Donnell
Illustrations: Erin DeGroot
Content Contributors: Information Security Practitioners

ISBN-13: 978-1-7344107-1-6

Our society has become dependent on the digital frontier to provide us better, faster, more interconnected capabilities. An entire generation has never experienced ordering a set of encyclopedias in order to do a book report or driven to a post office to purchase a stamp and mail a letter. These are the same people who see rotary phones and machines in amazement. The Internet has pushed new and amazing ways of exchanging information into our society and completely changed how we communicate with each other. Write a letter, buy a stamp, and walk the letter out to the mailbox. This is archaic to many people today.

As communication platforms change, the way that we store and access deeply personal, private and sensitive information is also changing. We rely on portable digital devices to manage our day-to-day lives and they allow us to do everything from finding a great restaurant to video chatting with our doctors. This access and convenience is unparalleled, building a reliance on digital capabilities that can force us to make very difficult decisions about the importance of our sensitive data versus the ease of access to important services. We've become dependent on companies to properly manage our data with the hope that they're doing everything possible to keep it safe and retain our trust.

This is why cybersecurity is so important. Along with ease-of-use and interconnectivity comes the risk of illegal access through many of the same connections we take for granted every day. Most of us blissfully go about our day surfing the web, ordering food on our smartphone or watching the latest video meme unaware of the battle that rages in the background to ensure these same services run without adversely affecting users. It's more than ensuring that a video will run without lag. Security is about protecting the very activities that these same users do every day as they pull up their bank account information or, God-forbid, connect to a public WIFI spot! People instinctively flock to the path of least resistance and will do things, that as cybersecurity professionals, leave us baffled. If you take a hard look, they are only doing these things because they don't know the implications of their actions. A typical user is oblivious to the dangers of each keystroke and swipe. This is also okay. This is also normal. We can't blame the user for being, well, a user. They don't have the same knowledge as we do and nor do I feel they should. Too much information regarding something you know very little about can be dangerous and create misplaced fears.

The role of cybersecurity professionals is to protect users from the miscreants of the web. Rest assured, there's no shortage of people trying to do bad deeds. Just as in the physical world, where there's something of value, there's someone who wants to take advantage of it or make it their own. Our job is to make every effort to protect users. It's one thing to create the illusion that all is grand online, and entirely different to actually help secure transactions and allow a user to feel confident when buying those fancy new kicks with one-day delivery. Most importantly, we have an obligation to protect users who rely on these systems of communication for life-altering situations: fighting back against oppressive regimes, standing up for civil liberties, ensuring their voices are heard and they receive the support they desperately need. Being interconnected is more than going shopping or sharing photos; being interconnected also means having the ability to create change, organize, and save lives.

Much like police officers doing their best to protect communities from crime while being short-handed and outgunned by savvy criminals, security professionals contend with similar circumstances in the digital world every day. It's been reported that there's a need for nearly 3 million cybersecurity professionals to fill the current and future jobs responsible for securing our connected world. It is estimated that in 2019, there are over 14 billion connected devices. The gravity of the need for more cyber-defenders has become more serious than ever before. Bad actors have clearly taken notice by leveraging these same connected devices and using them to commit crimes and take advantage of unsuspecting users.

I want to thank you for stepping into the fight by reading this book. The need for people who have a clear desire to protect and defend our cyber systems has never been more critical. Each day there is another article about a new breach, hack or data leak, in many cases stemming from administrative tasks that unfortunately were overlooked and might have prevented the attack. Sometimes, a simple user error like opening a document laced with malware or clicking on a malicious link is all it takes for very bad things to occur. We are human and mistakes happen. It's critically important that as you advance in your career that you don't become jaded and begin blaming the user. As I mentioned earlier, the user is simply doing what most non-tech people would do and mistakes will happen. It's our job to do our very best to protect them and have empathy when they make a mistake. We need to be the advocates for security awareness through the use of our technical knowledge and expertise. Much like we try to shield children from negativity, we must work to shield our users so they can focus on what matters most to them and be as productive as possible using the latest technology advancements available. In essence, you are an unsung hero working tirelessly in the background to keep things running smoothly. I encourage you to embrace this role and celebrate the fact that every day you are making the cyber world a safer place for everyone.

Information security is a diverse field covering everything from networks and applications to open source intelligence to threat analytics. As you read this book, you'll find a wealth of information and topics that will not only prepare you for taking the Security+ certification exam but it will also help to shape the direction within security you would like to go. You may even identify with an area that you would like to specialize in. Yes, it's perfectly fine to be a specialist. In fact, every mentor I've had has recommended becoming a specialist to me because of how vast our field is. Take a moment to look deeper into the topics that really pique your interest and see if it's something you might want to dive further into. This field touches so many areas of business, society, and culture that you will never run out of challenges to tackle. You've taken a great step by preparing for the Security+ exam as it will give you deeper insight into some of the very challenges you'll face. You will also need to have the tenacity to go out and gather knowledge and experiences based on industry trends and evolving tactics, techniques, and procedures of bad actors. If you stay sharp and practice your skills, I have no doubt you'll find an exciting career and enjoy the satisfaction of knowing that you contribute to making the cyber world a safer, more productive place.

Rey Bango
Security Advocate
Microsoft Corporation

TABLE OF CONTENTS

THREATS & ATTACKS

INDICATORS OF COMPROMISE: COMMON THREATS

VIRUSES

A virus is a form of malware that targets all computers on a network. Viruses propagate by infecting targeted files to affect all of the connected drives on a computer system. After successfully infecting a system, the virus is capable of collecting data, harvesting credentials, sending emails, crippling computers and much more.

CRYPTO-MALWARE

Once a system is backdoored, cryptocurrency-mining software can be transferred to the compromised system. By design, crypto-malware runs as a background process in order to evade antivirus detection. Hackers can generate a constant flow of illicit revenue by utilizing many compromised computers simultaneously.

As malware consumes CPU and GPU power, performance on infected systems decreases. Consequently, this can also decrease the life cycle of the computer.

RANSOMWARE

Ransomware is a type of malware that encrypts files on a system then offers to decrypt them in exchange for money - typically in the form of cryptocurrency.

Ransomware such as WannaCry or NotPetya utilize the EternalBlue exploit leaked by the Shadow Brokers hacking group in 2017. When executed, the EternalBlue exploit will allow the attacker's shellcode to generate a system level shell which can be leveraged to install the WannaCry ransomware, join the system to a large botnet, or even use the victim machine as a pivot point into a network.

The NotPetya ransomware acts like a worm in that it is self-spreading. After successfully infecting a victim, the malware scans the remaining network for potentially vulnerable Windows machines. Once machines are detected, the malware will attempt to exploit each machine with the EternalBlue exploit. NotPetya can also leverage a tool called Mimikatz to dump network administration passwords before using PsExec and WMIC tools to pivot to networked machines.

KEYLOGGER

Hardware and software keylogging is the action of recording user keystrokes and relaying them to the attacker. This effective method is used to capture and obtain sensitive user credentials.

Keylogger software no longer serves a single purpose, but is versatile in nature. Many commonly used keyloggers are also capable of installing files, destroying data, backdooring system files, and much more.

SPYWARE

Software created with the intent of capturing and exfiltrating data about the victim is known as spyware. Spyware evades antivirus detection while capturing metadata and sending it to the attacker. It can originate in malware such as password stealers, banking trojans, info stealers, and keyloggers. They can even exploit web browser vulnerabilities to intercept web form data and saved passwords.

BACKDOOR

A backdoor grants access to a computer system by bypassing normal authentication. They usually result from infection by a virus or trojan. With backdoor access, a hacker can privilege escalate and manipulate a compromised computer as if sitting in front of it.

INDICATORS OF COMPROMISE: LESS COMMON

WORM

A worm is malware that self-replicates and pivots across networked computers. Peer-to-peer and client/server networks are vulnerable to worm migration. Worms will often install bot clients on vulnerable machines.

TROJAN

A trojan or trojan horse is malware embedded in an otherwise "normal" file, like a document, image, or program. When a user accesses the seemingly innocuous file, the embedded malware executes, as well. Once executed, the legitimate looking file creates a local backdoor which provides an attacker remote access to the victim's filesystem.

ROOTKIT

Rootkits can install themselves at the core of operating systems and computer hardware. Some rootkits, whether installed via user-clicked links from spam email, ftp, or drive by download, can install at the firmware level.

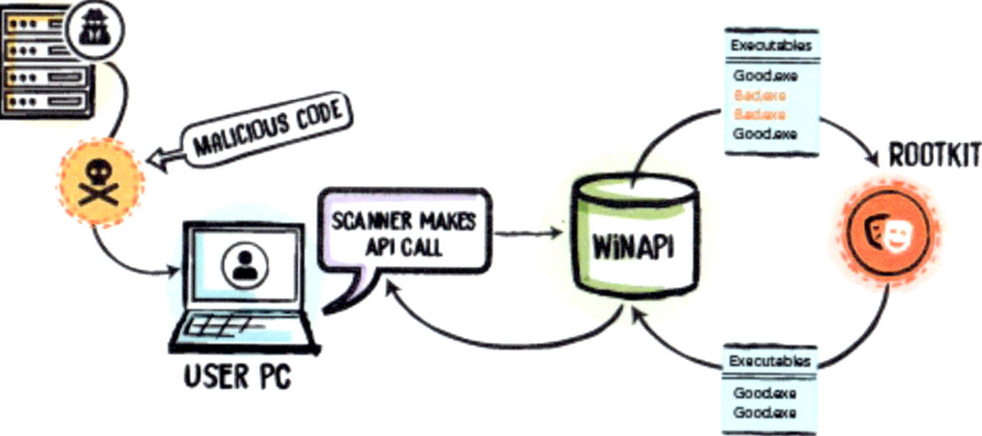

Firmware rootkits are boot and antivirus persistent. Even if detected and removed by antivirus methods, firmware rootkits will simply reinstall at the next boot. Some forms have the ability to persist through OS reinstallation, as well.

Router rootkits will modify network configurations as well as route all traffic to malicious content. All systems, including cell phones, networking devices, and SCADA systems, are susceptible to rootkit installation.

ADWARE

Adware is installed to a computer when a user clicks compromising links or browses to a malicious website. This malware typically collects data, redirects requests to advertising websites, and tracks system activity surreptitiously. Unwanted pop-ups and unexpected redirects are both indications of possible Adware infection.

BOTS

A bot is a client that is installed on a compromised system and that executes commands issued by a Command & Control (C2) server. These clients are attached to various adware, malware, and spyware. Many bots are DDoS clients that collectively send data to a target IP address. Other bots are capable of keylogging, relaying spam, capturing packets, and opening backdoors.

RAT

A RAT ("remote access trojan") is a backdoor method for taking full control of a computer. A RAT operates with the same level of user permissions as the user who downloaded it. Once a RAT has been successfully installed, an attacker can control a system remotely with the same levels of access (and potential for damage) as a human user. This includes the ability to harvest data and install further malware.

LOGIC BOMB

A logic bomb is a tactical piece of code/malware inserted into otherwise functional software. The logic bomb is designed to execute when predetermined conditions are met. Triggering events include specific time/date conditions, on the execution of a program, or failure of a user responding to the program. This malware is employed primarily for destructive purposes. For instance, it might delete a database, overwrite hard drives, or release sensitive data.

TYPES OF ATTACKS

SOCIAL ENGINEERING

Rather than spending weeks or months gaining initial access via a vulnerable computer system, a hacker may be able to compromise a machine by simply manipulating a person into divulging secure information. Social engineering relies on two general truisms of human behavior: 1) People want to be helpful. 2) People want to avoid uncomfortable situations. Therefore, people are often considered the weakest link in regards to an organization's cyber security.

For instance, one might gain access to a front desk clerk's work space or by impersonating IT personnel. The slightest information about standard practices, including password restrictions, or naming conventions for emails and machines, can be valuable to a potential hacker. Attaining names and addresses of domain controllers, workstations, routers, etc. would allow further enumeration of a network.

APPLICATION / SERVICE ATTACKS

Almost all businesses use some sort of web application. In an enterprise environment it is common to find intranet accessible web applications that are hosted locally on company servers. These web applications, if not tested properly and routinely updated, can have many bugs. Bugs in unmaintained web applications can be exploited to gain shells on production servers. Common web app attack methods include injection, cross-site scripting, broken authentication and session management, and more. When a production server web application is exploited, the attacker gains a foothold on the local area network which can contain sensitive databases full of employee/customer stored data.

WIRELESS ATTACKS

Wireless networking provides many benefits, but also introduces additional security concerns. Wireless network security has evolved from WEP (Wired Equivalent Privacy) to WPA (WiFi Protected Access) to WPA2 (WiFi Protected Access II) with each successive protocol providing increased security.

Open source tools such as the Aircrack suite are capable of attacking WEP and WPA networks. WEP networks are easily penetrated by injecting ARP traffic into the network while simultaneously capturing all the traffic. Given sufficient traffic, Aircrack can crack WEP encryption within minutes.

WPA is harder to target than WEP and is less likely to be successfully cracked. While the WPA protocol itself has been attacked successfully, many WPA and WPA2 routers are susceptible to indirect methods of access through a default-enabled feature, WPS ("WiFi Protected Setup").

WPS is a user-friendly shortcut for joining devices to a WPA or WPA2 network. A physical button located on the router initiates the WPS process. After initiation, the router opens itself to requests from devices available to connect via an 8-digit PIN. WPS-enabled routers are thus susceptible to brute force attacks which can be successful in as little as a few hours.

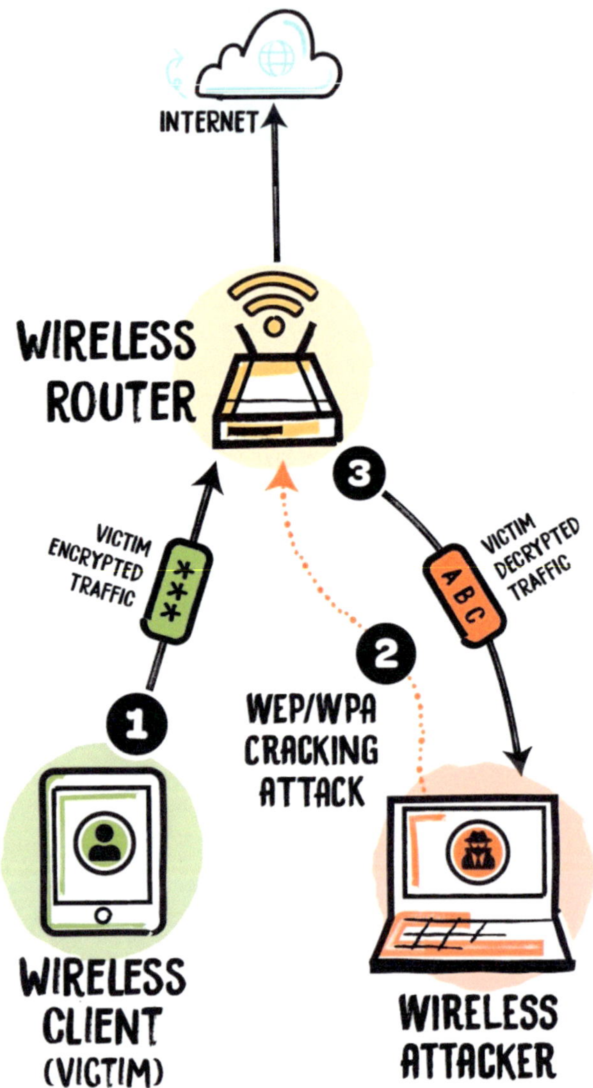

CRYPTO ATTACKS

Cryptanalysis and crypto attacks result in ciphertext being decrypted to plaintext. The goal of a crypto attack is to find the secret decryption key for a ciphertext. The complexity and volume of possible decryption solutions for a given ciphertext means that many attack methods exist and all of them require computer processing power to accomplish. Under given circumstances, if part of the plaintext is known, then the ciphertext could potentially be decrypted.

When KPA (Known Plaintext Attack) attacks are carried out successfully, the secret key or code book (a table with words or phrases that have correlated strings) could be gathered.

In a BFA (Brute Force Attack), the attacker guesses all possible keys in attempting to guess the correct one. BFA attacks are inefficient and time consuming.

In a MITM (Man in the Middle) scenario, the attacker must be in the middle of a key exchange prior to communication occurring between two hosts. If the attacker intercepts the first host's key and replaces it with their own, they can then control all traffic while maintaining the illusion that traffic is secure and legitimate. The attacker must re-encrypt traffic before sending the data to the receiving host in order to maintain communication. This process of decryption and encryption is a recurring cycle until the attacker obtains the entirety of secret data.

THREAT ACTORS & ATTRIBUTION

TYPES OF ACTORS

SCRIPT KIDDIES: A script kiddie is a person, generally of low technical ability or knowledge, who uses existing hacking tools to attack computers maliciously. Disregarding how the tools work or how to code them, they use premade exploits to quickly gain access. They are motivated solely to achieve their desired outcome, not in developing technical skill or knowledge. As they are not seeking sophisticated solutions, they often apply openly available applications (Low Orbit Ion Cannon, for instance) to a multitude of attacks regardless of whether the tool is technically appropriate. Their 'Hail Mary' usage of these attacks can result in computer services crashing in a (sometimes unintended) DoS attack.

HACKTIVIST: An individual or group which breaks into computer systems with the intent of furthering a political or social agenda. They will attack public websites and private networks of governments, officials, corporations, and religious organizations which promote or represent perceived social or political injustices. Defacing websites, breaking into and leaking information, and DDOS attacks are typical hallmarks of hacktivism. Hacktivists can be funded by governments looking to stir opposition among rival nations or states.

ORGANIZED CRIME: Organized crime is the cohesive functionality of active threats. Crime-as-a-service creates a scenario where the attacker merely purchases hacking services or malware, and thus might remain entirely ignorant of the hacking tools and coding used in an attack. Purchasing malware allows attackers to plug-and-play exploits with previously developed tools. This enables attackers to mass exploit or target a single entity.

The availability of cybercriminal services and products is a catalyst for many underground sales points. The dark web has been known to host the sale of weapons, drugs, and many other illegal items. Encryption and obfuscation are commonly applied to communications within these alternative markets.

NATION STATES / APT: Nation states are interested in political and economic espionage. They seek to obtain state secrets, intellectual property, and the personal information of government employees. The process of orchestrating a sophisticated, timely attack coins the term advanced persistent threat. Whether it's a group of privately-funded individuals or a large state actor, gaining persistence on the target enables them to consistently exfiltrate data and monitor the target. Another hallmark of APTs is their relentless pursuit of a target. Even if defenders successfully stop an attack, APTs will simply try again after a time.

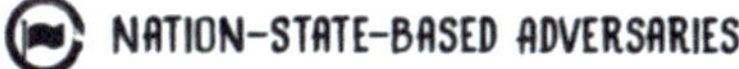

INSIDERS: Companies spend significant resources on protecting their network from outside threats. Mitigating potential insider threats should be given equal attention. An insider threat is a person or people within an organization who act maliciously against it. This could include employees, former employees, or contractors with access to company systems. "Least privilege" is a principle implemented by systems administrators which seeks to limit user access to network resources to the lowest level needed by that user. This principle helps to avoid instances in which a user (or worse, every user) has unrestricted access to network resources. Giving administrative privileges improperly may result in the release or destruction of sensitive data.

 Threats lie on all sides of every organization. At times, threats are highly visible and at other times unpredictable and difficult to discern. Competitors are a large threat when their agenda is to take out the competition using illegal hacking. Sabotage and theft of data may result from these attacks when carried out successfully. A competitor might infiltrate a company using outside actors, or might use existing insiders from the target organization. The end goal is to inhibit the competition's growth.

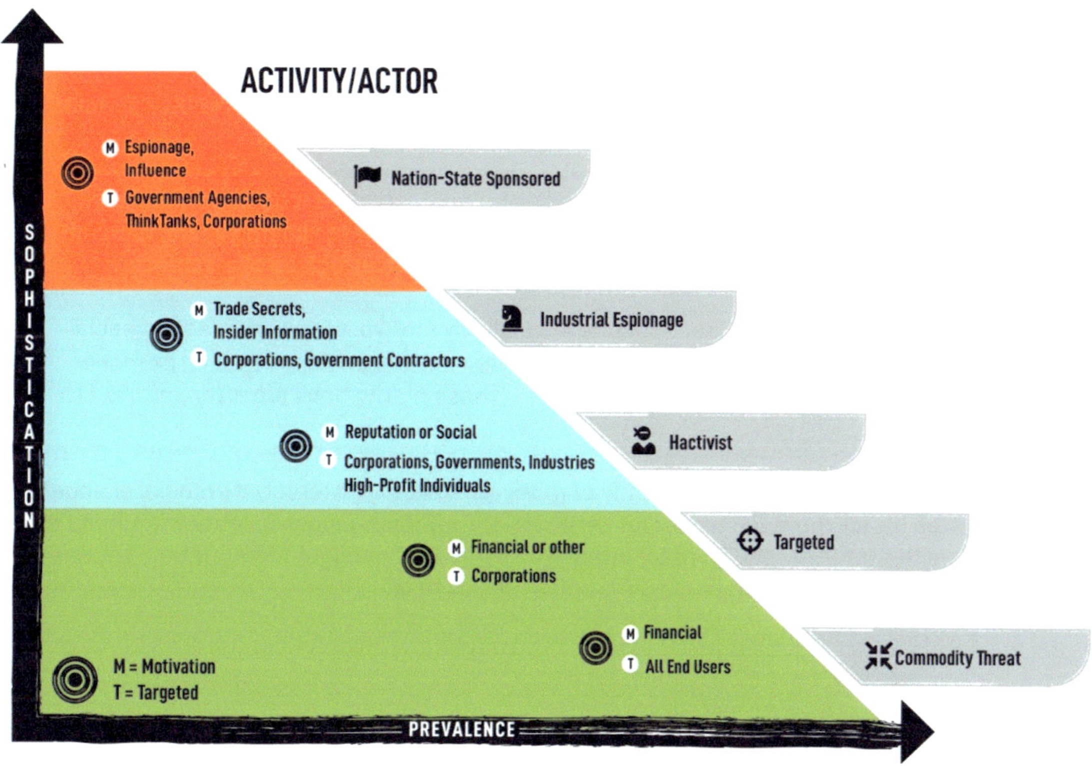

INTERNAL / EXTERNAL: Sometimes a great threat lies on the inside of a network. Insider threats will have very good understanding of the topology of the network they are attacking. The threat may be an employee who has intentionally taken the time to fully understand the inner workings and cohesion of a network so that they can identify and target systems strategically.

While an external threat such as a nation state may lack insider intel, it would be fully capable of deploying a sophisticated attack due to an abundance of time and resources.

LEVEL OF SOPHISTICATION: Successful attacks can be launched by individuals and teams of widely differing ability. The outcome of a successful attack is the same regardless of whether executed by a highly sophisticated, knowledgeable team, or a lone actor with limited understanding of the tools and processes employed.

For instance, thousands of nuclear reactors were crippled by malware which was developed and deployed by a team of sophisticated, highly knowledgeable actors working on behalf of a nation state. On the other hand, someone with little fundamental knowledge but high persistence to gain access to an organization might get lucky deploying public tools against vulnerable machines on a network. For example, there has been an explosion in ransomware infections on business networks over the last 5 years. In 2018, a state government network in Colorado, USA, was infected by the SamSam ransomware, and one of the world's largest aluminum producers, Norsk Hydro in Norway was partially shut down and had to switch to manual operations in 2019.

RESOURCES / FUNDING: Script kiddies use open source software and have very little funding, if any at all. Generally, they are motivated by making a name for themselves.

Contrasting script kiddies, nation states possess seemingly unlimited resources. Abundant funding provides access to highly skilled hackers who are devoted to accomplishing incredibly sophisticated attacks. Hackers may be hired by governments to attack foreign targets.

INTENT / MOTIVATION: An attacker might be intrinsically motivated toward creating social change, pursuing a political agenda, or fun. Alternatively, external incentives such as money might motivate an actor to attack a specified target. These motivations are often referred to as 'for fun or profit'.

The intent of an attack varies independently of motivation. Actors interested in social change might launch an attack intended to disrupt services, or they could launch an attack intended to exfiltrate sensitive information. An actor interested in having fun might merely intend to peer behind a locked door or cause as much destruction as possible.

VULNERABILITIES

PENTESTING CONCEPTS: PLANNING

BLACK BOX VS WHITE BOX VS GRAY BOX

Black Box pentesting is a simulation of a real world attack. To best simulate real world attack conditions, the attacker is given no prior knowledge of the network or infrastructure being tested. This form of pentesting is disadvantageous to the attacker, but provides the client an accurate assessment for the potential success of an uninformed, malicious attacker.

White Box penetration tests are facilitated by an informed attacker. This allows for a deeper and more thorough assessment of a network in a given period of time. In a white box pentest, the tester is given as much info as necessary to accomplish a thorough test, including application source code, network ranges of workstations and servers, and more. Providing this information from the outset allows the tester to focus their allotted time on exploiting identified vulnerabilities instead of scanning to discover them.

Gray Box testing lies somewhere in the middle of these tests. Testers are given some information to help them complete a more effective test but not enough that they are able to own the portion of the network that needs a more black box style test.

PENTESTING VS VULNERABILITY SCANNING

Pentesting and vulnerability scanning are entirely different but equally necessary aspects of securing a network. Vulnerability scans attempts to identify publicly known vulnerabilities in a network. Penetration testing is the exploitation of vulnerabilities identified on the network.

While vulnerability scanning takes a large amount of time, it is the necessary foundation of all further testing. Note that scanning of a subnet can be entirely automated. Free scanners such as nessus, nmap, nikto, sparta, and openvas are commonly used to scan for vulnerabilities prior to pentesting.

Although the scans can be fully automated, the exploitation phase should be done manually. While automated fuzzing applications save time and effort, manually fuzzing ensures that each system in a scan is thoroughly examined. Automated tests cannot analyze a scenario and adapt their methodology but rather only give a fixed result. Manual tests ensure that the expert pentester will give full focus to each system.

PENTESTING CONCEPTS: INITIAL ACCESS

ACTIVE / PASSIVE RECONNAISSANCE

Active and Passive reconnaissance have two very different standards.

Active recon engages the target and its systems directly while expecting to leave a footprint. This would involve scanning each system for open ports and services running as well as enumerating DNS records. Each query is easily detectable but far more informative than the slower and less aggressive passive recon.

Passive recon gathers information via third party resources such as social media or google while never making direct requests or scans to the client. Gathering information on the target may be accomplished by searching discarded documents for usernames and emails, eavesdropping on employee conversations, utilizing whois, and packet sniffing.

PIVOT

Pivoting is the act of migrating from one victim computer to another. Pivoting is an essential action for attacker to reach their final destination in a network. Whether they utilize operating system exploits or dump administrative credentials, pivoting is a very effective way to take control of a network. This also allows the attacker to mine data from computers that may have more accessibility or sensitive information than others. For example, if a server has VPN connectivity to a segregated network that all others don't, the attacker can route all of their probing traffic through the exploited server to access an entirely new subnet of computers.

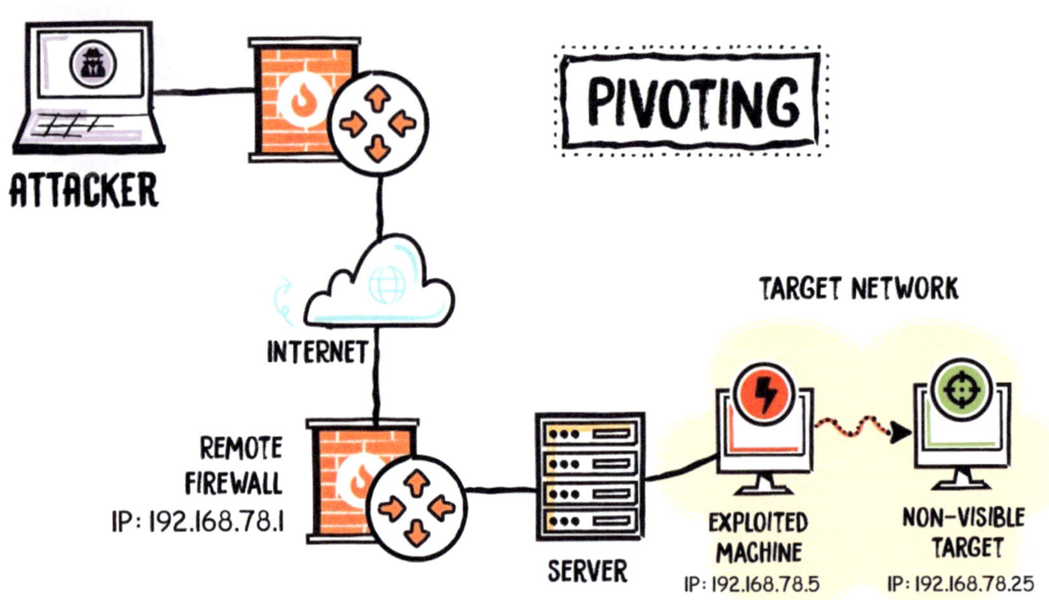

INITIAL EXPLOITATION

The initial exploitation of a system is the primary entry into a networked computer. It is vital that this system and its services do not crash or shutdown inhibiting the attacker from maintaining access. Utilizing Windows and Linux exploits is a common tactic used by hackers to gain initial access to a computer. Once reconnaissance is complete, the attacker has obtained a detailed table of open ports and services running on each machine. Depending on the version of these services there may or may not be public exploits that are available to attack the machine. The risk of utilizing these exploits is that when they go untested prior to a pentest; they are able to crash vital services running on the target machine. If the exploit fails on the first attempt of gaining a reverse shell, the service may crash therefore inhibiting another attempt until the computer is rebooted or the service is restarted. This invasive method should be an attacker's last resort to gaining access. A less risky approach may be spear phishing, drive by downloading, or infecting a client's web browser with a BEEF hook and social engineering.

PERSISTENCE

One of the most important aspects of gaining access to a network is maintaining access. If much time and effort is spent gaining initial access then the acquired shell is very valuable. Sometimes the victim may only click an email or open a downloaded document one time. This means that the system exploited must be further enumerated and exploited if your shell does not have system or root access. If the attacker only has a user level shell then they may not be able to maintain persistence on a machine as easily as a higher privilege shell. This is because a standard user shell may not have administrative rights which enables a user to stop / start services, create startup scripts, and modify a computer's registry. However, if the attacker does gain a high privilege shell, then it would be easy to create any of these as well as modify or replace system files with a malicious version. Persistence is important because if you wish to maintain access through reboots, logouts, or crashes then your method of persistence must create a call back to your CNC every time one of these events occur otherwise your shell will die.

PRIVILEGE ESCALATION

Many times when a machine is compromised via a higher level exploitation method such as phishing or downloading a malicious file, a user level shell is created. Unless the user that executed the malicious file is part of the administrative group on the domain then the shell will have very few rights. At this point the attacker must find some route to escalate privileges if they wish to obtain administrative rights on the machine. The attacker can then search for services or files with weak permissions, potential operating system vulnerabilities, or some way to generate a system level shell. It is also possible to scan other machines on the network for vulnerabilities that will produce a system level shell once exploited.

VULNERABILITY SCANNING CONCEPTS

PASSIVE SECURITY CONTROL TESTING

Passive testing entails much less network intensive methods but results in a less in-depth collection of data. By analyzing existing traffic flow, passive scanners can deduce much about the systems communicating on the network, such as the operating system, application types, and version information. Passive scanners utilize the information that computer systems expose about themselves in normal communication. A disadvantage to passive testing is if an application or service is not utilized commonly, the scanner may not detect it whatsoever.

VULNERABILITY IDENTIFICATION

Vulnerability identification is the next stage after a vulnerability scan. Once a full scan is complete, the pentester is then tasked with cross referencing running services and web vulnerabilities with public exploit databases. Sometimes scanners such as nikto or nessus will display links to CVE descriptions or even exploit code. Dictating the version of services running on open ports is vital to the process of identifying potential vulnerabilities. Outdated and unpatched applications are common points of compromise that likely have available exploits in online exploit databases, such as exploit-db.

LACK OF SECURITY CONTROLS & COMMON MISCONFIGURATIONS

If a web application does not have configured security controls, an outsider can directly access the application. Whether it be HTA login forms or white/black listing of IP addresses if either security controls are absent then the backends of each server are accessible by outsiders leaving ports scannable and services exploitable.

INTRUSIVE VS NON-INTRUSIVE

Port scanning, scanning for missing updates or patches, and scanning registries for values are examples of non-intrusive scanning. Probing around for vulnerabilities is a non-intrusive way to inspect the state of security for a computer system. When a scanner attempts to exploit these vulnerabilities, it then becomes intrusive. SQLMap is an intrusive scanner that will look for injection vulnerabilities and exploit them. It is vital to obtain the client's permission before running either scans with an emphasis on intrusive scans due to the potential to crash services unintentionally.

CREDENTIALS VS NON-CREDENTIALED

Credentialed scans will scan networked machines with specific services and accounts enabled. This scan goes deeper into potential problematic applications, file permissions, and registry issues that are not visible at the network level. Non-credentialed scans are more surface level because they only scan networking services on each host. They reveal open ports and running services on each host.

POTENTIAL IMPACT OF VULNERABILITIES

Potential of impact with regard to a vulnerability is proportional to the probability of success of the exploit. This drawing represents the various probability vs impact scenarios.

LOW PROBABILITY & LOW IMPACT

- Out of date flash on a managed desktop host
- XSS for out of date technology on internal facing development website
- DoS in packing software used to pack internal only executables - CVE-2019-1010048

LOW PROBABILITY & HIGH IMPACT

- Exposed development credentials on an internal file share from 2016
- Internal only domain controller unpatched for three business cycles
- Rogue process reading memory from microprocessors - CVE-2017-5753, CVE-2017-5754

HIGH PROBABILITY & LOW IMPACT

- XSS in internet facing development website - CVE-2012-6644
- DoS against company sales websites
- Improper permissions for old web framework allows session IDs to be read - CVE-2012-2760

HIGH PROBABILITY & HIGH IMPACT

- Web facing production administrator portal for customer services
- Default credentials for critical routers
- Managed hosts with SMB open to the internet - CVE-2017-0144

When a vulnerability is found inside of a network it is not uncommon for the systems administrator to be lackadaisical about it due to the resource not being internet accessible. It is not assumed that these vulnerabilities will be taken advantage of. Whether it be a missing Windows patch, an outdated Apache server, or a misconfigured WSUS server, if an attacker is able to gain access to the network, then traversing to vital resources may be imminent. In the case that the administrator did not configure WSUS to push updates regularly, every machine on the network may be missing patches and be vulnerable to recently published exploits. However, when a vulnerability is found on a machine facing the public internet, it is usually addressed immediately due to the critical nature of the situation. With all subnets being constantly scanned by threat actors, it's only a matter of time before vulnerable machines facing the internet are exploited.

TECHNOLOGIES: SECURITY SUPPORT

FIREWALL, ROUTER, SWITCH INSTALLATION & CONFIGURATION

ACL or Access Control Lists are filters which routers and some switches use to allow or deny traffic on a given interface. ACL lists use stateless inspection, meaning they analyze each packet to see if there is a stream of data to which it belongs or if it's a rogue packet. While a network device is inspecting traffic passing through, if the data meets a criteria given by the ACL list it will either permit or prohibit the data. Although ACL lists are not as thorough as firewalls, they are great for network interfaces that have mass traffic being routed through which helps avoid rate-limiting. They are also good for network perimeter routers that block global routing tables.

There are different types of firewalls, like application-based firewalls and network-based firewalls. An application firewall is a form of firewall that controls input, output, and/or access from, to, or by an application or service. It operates by monitoring and potentially blocking the input, output, or system service calls that do not meet the configured policy of the firewall. Network firewalls guard an internal computer network against malicious access from the outside, such as malware-infected websites or vulnerable open network ports.

Stateful and Stateless packet inspection are methods used by firewalls. Stateless packet filtering is much less effective than stateful filtering due to packets being analyzed by source/destination and other static values. Stateless firewalls are 'unaware' of any previous data stream that a packet may or may not be a part of. They use ACL's to filter packets whereas stateful firewalls analyze all previous data streams end to end and are able to implement IPSec functions such as tunnels and encryption. Stateful firewalls are capable of dictating the stage of a TCP connection, which means they are able to tell if the MTU (maximum transmission unit) has changed or if a packet has fragmented.

The implicit deny security stance treats everything not specified as whitelisted or permitted as suspicious. An example of this is network boundary rules that drop all traffic except for whitelisted IP addresses and service ports while blocking all others.

Anti-spoofing is a technique used to identify and drop packets that have falsified source addresses. This is a common attack method used for DoS attacks. It is also used when trying to spoof traffic which needs to look like it is coming from a whitelisted IP address. This is how an attacker may be able to exploit an unreachable target.

Switch ethernet LANs are vulnerable to MAC address spoofing and Layer 2 denial-of-service attacks therefore port security refers to the prevention and detection of these attacks. On some devices, ports can be categorized as trusted or untrusted. Routing loops are another potential threat to networks if not configured properly. Looping causes packets to continue to route in an endless circle rather than reaching their end destination. Split-horizon, route poisoning, and hold-down are some mechanisms which help prevent looping. Flood guard may be used by switches that receive MAC flood attacks. When a flood occurs, it fills the internal table where MAC addresses are stored which in turn will eat all the memory on the switch if too many MACs are overloading the table. Flood guard will prevent this by limiting the amount of memory used to store MACs, shutting the port down until resolved by the administrator, and restrict updates for the port.

AP & VPN INSTALLATION & CONFIGURATION

SSID or Service Set Identifier is the name associated with an 802.11 wireless network or hotspot. It is assigned in the access point configuration menu. In the configuration menu, MAC filtering may be enabled. When enabled, administration of MAC addresses by whitelisting and blacklisting is possible. Only whitelisted MACs are able to connect to the access point.

The performance of WiFi networks greatly depends on the radio signal strength. Updated routers are capable of broadcasting 2.4GHz and 5GHz network signals. Network speed is dependent on the distance between the router and connecting device. Obstruction of a signal by walls and other items will prohibit a device from effectively utilizing the speed of a 5GHz network whereas 2.4GHz travels through them more easily. The utilization of high quality antennas and strategic placement of network devices will greatly improve the overall functionality of wireless networks.

Stand-Alone Access Points (Thick) and Controller based Access Points (Thin) are options for an enterprise environment. Controller based access points communicate to a centralized management system where configuration, encryption, update and policy settings are configured via a centralized controller. Stand-alone APs are exactly as they sound. They are configured from a PC and operate in communication with other APs.

Remote access and Site-to-Site VPNs are similar in nature but are utilized differently in business environments. Remote access VPNs connect single users to remote networks while encrypting data prior to being sent over the public network, and then decrypting the data once it reaches its destination. Site-to-Site encrypt and decrypt similarly, but their purpose is to connect entire networks together via encrypted tunnels causing seamless communication between remote resources.

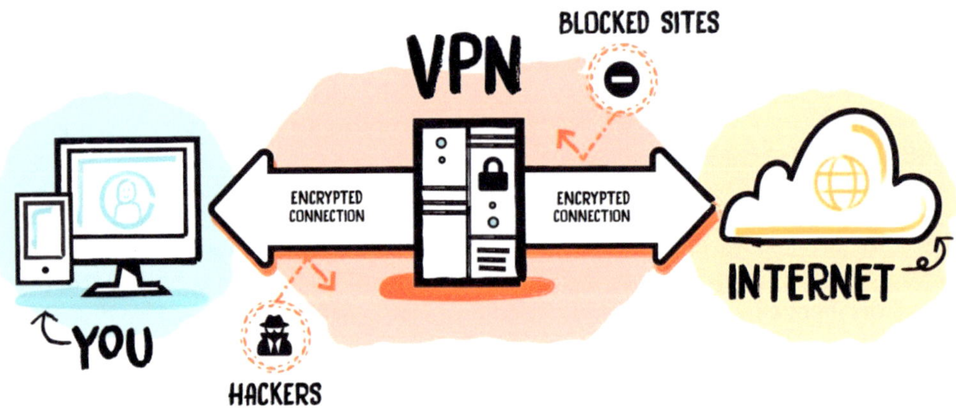

IPSec is a set of protocols commonly used by VPNs which authenticates and encrypts every packet sent over a network. IPSec can be run in either tunnel mode or transport mode. With tunnel mode, the entire original IP packet is protected by IPSec. This means IPSec wraps the original packet, encrypts it, adds a new IP header and sends it to the other side of the VPN tunnel (IPSec peer). Tunnel mode is used to encrypt traffic between secure IPSec Gateways, for example two Cisco routers connected over the Internet via IPSec VPN. Transport mode encapsulation retains the original IP header. Therefore, when transport mode is used, the IP header reflects the original source and destination of the packet. Transport is most often used in a host-to-host scenario, where the data endpoints and the security endpoints are the same. AH (Authentication Header) provides anti-replay, data integrity and data origin authentication features, but does not support encryption. However, ESP (Encapsulation Security Payload) support authentication, data integrity, anti-replay and data encryption features. Therefore, ESP is a more robust protocol. TLS (Transport Layer Security) is an alternative encryption method to IPSec used by VPNs at layer 4 of the OSI model rather than layer 2. TLS is used for securing application data traveling across the tunnel from web server to client.

A split tunnel VPN encrypts and sends specified data such as Canvas, Outlook, or other services over the VPN tunnel but routes all other non-important traffic such as Facebook, YouTube, and Yahoo without encrypting or sending it over the VPN. Split tunneling offers the potential threat of traffic not traveling over the VPN if misconfigured. Alternatively, a full tunnel routes all data streams over the VPN. Considering VPN bandwidth will help avoid bottlenecking the tunnel.

SIEM & NIPS / NIDS INSTALLATION & CONFIGURATION

SIEM (Security Information and Event Management) products unify information management and event management by providing real-time analysis of security alerts generated by software applications and network devices. Aggregation is the process of reducing the volume of event data by consolidating duplicate records and reports of event data then moving the data to a common repository. Event correlation is an essential part of any SIEM solution. It aggregates and analyzes log data from network applications, systems, and devices making it possible to discover security threats and malicious patterns of behaviors that otherwise go unnoticed and can lead to compromise or data loss. Custom rules can be configured to trigger alerts when specific conditions are met. Various log sources, such as network devices, security devices, servers and antivirus, provide data for thresholds to be triggered based on user authentication rules, attacks detected and infections detected. Thresholds can be configured to trigger alerts based on the quantity of occurrences of events.

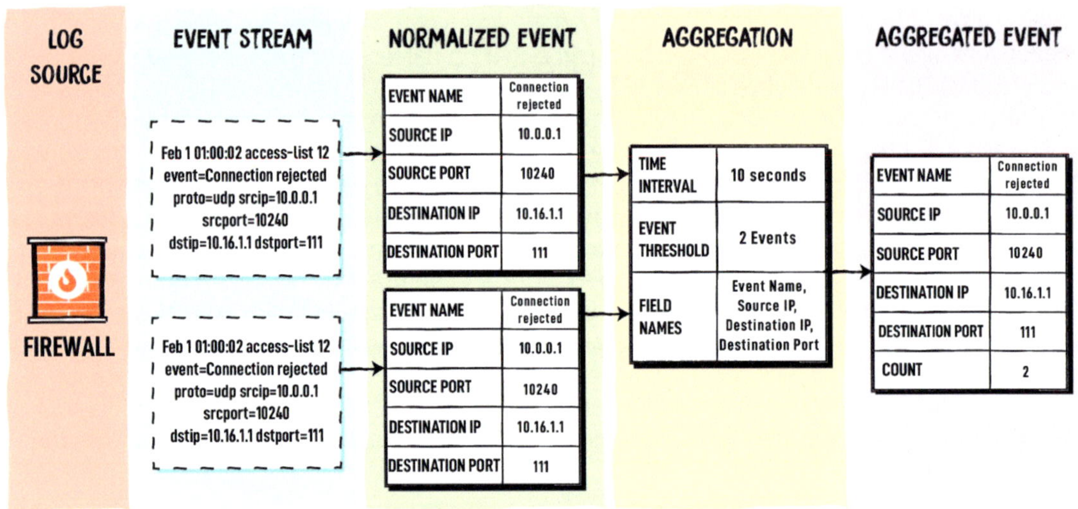

HOW EVENT AGGREGATION WORKS

Signature based IDS (Intrusion Detection System) refers to the detection of attacks by looking for specific patterns, such as byte sequences in network traffic, or known malicious instruction sequences used by malware. Signature based IDS are only as effective as the signatures it is monitoring for. Much like anti-virus software, it must be updated regularly to stay up to date with varying signatures. Heuristic intrusion detection looks for behavior that is out of the ordinary. A baseline of normal activity is taken on the network then compared to all other traffic. When an event occurs that is out of the ordinary, a summary of the alert is produced. Having a trained professional to review either signature or heuristic events is essential to recovery. Signature based IDS can be fooled if an attacker is able to bypass the filtering of a signature using polymorphic code or obfuscated/spoofed headers. Heuristic detection systems produce false positives more frequently and are more hardware intensive.

IDS are passive systems that scan traffic and report back on threats. IPS (Intrusion Prevention System) are placed inline or in the direct communication path between source and destination, actively analyzing and taking automated actions on all traffic that enters the network. Dropping the malicious packets, blocking traffic from the source address, resetting the connection, and sending an alarm to the administrator are actions an IPS may take based on given rules.

ENCRYPTION & MAIL GATEWAY INSTALLATION & CONFIGURATION

Due to unsolicited emails, organizations must implement mail gateways to perform checking for spam, data integrity, and data coming through is intended for recipients on the inside of a network. Inbound and outbound communications are monitored for unsolicited spam, advertisements, and phishing attempts. If a phishing attempt is detected, the mail gateway may diffuse the attempt by changing the link or blocking the email at the gateway. Anti-virus and DLP (Data Loss Prevention) may be included on the mail gateway. The anti-virus will scan all inbound and outbound attachments for malicious files then block them. DLP will filter emails of personal or proprietary information to prevent leaking any sensitive data.

Methods such as configuring a whitelist and blacklist by allowing messages only from known locations while blocking all other emails, and verifying that emails conform to specific SMTP standards and dropping all other non-RFC compliant may help prevent spam from entering a mail server. Reverse DNS on inbound email traffic will help compare the sender and IP address of where the message was sent from and the domain name IP address. If they do not match, then the email can be discarded as spam. Intentionally slowing mail delivery is known as Tarpitting. This may prevent a spammer from continuing to send mail due to the delay of delivery.

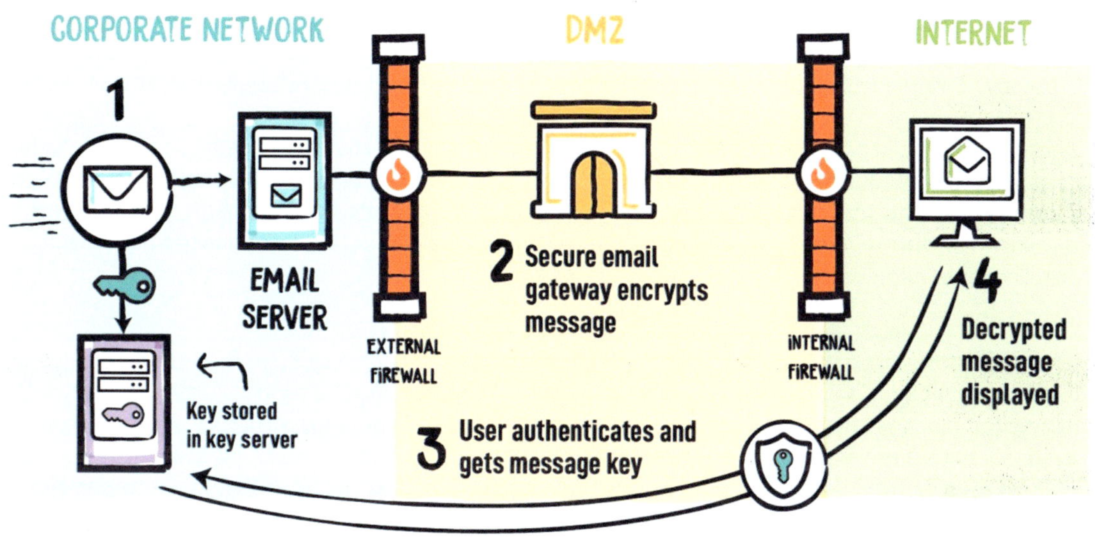

Mail being sent and received can easily be intercepted. Unless the email is encrypted, the traversing mail would be in clear text. When sensitive information is included in a message, the information should be encrypted prior to or at the gateway. The gateway can be configured to verify that sensitive information is encrypted. If it is not already encrypted, the gateway will encrypt the message. If the message was encrypted by the client, then the gateway will confirm this and allow the message to be sent.

TECHNOLOGIES: SECURITY POSTURE ASSESSMENT

NETWORK, COMPLIANCE, & VULNERABILITY SCANNERS

A protocol analyzer such as Wireshark is used to passively capture packets. Once captured, the information inside the packet may be inspected in plain text. Captures can be taken on a wireless or wired network. Capturing packets allows you to be able to look for traffic patterns and what protocols are going across your network. This will help identify unusual or malicious traffic. Using analytics to examine a large capture may help identify complex issues.

Using a network scanner is an active way to gain a bird's eye view of a network. Network scanners such as Nmap or Zenmap are used to determine what services may be running on a remote device such as IRC, ftp, or http as well as the operating system on the host. You can specify a single IP address to scan or a range of IPs. This helps to discover known or unprecedented devices on a network. Network scanners help expose potential vulnerabilities as well as give a network topology. Zenmap will display a graph of the scanned network range to visually map the network.

Banner grabbing is a term that describes the process of parsing the response of a request made to an open port with a service running. It is used to gain information about a computer system on a network and the services running on its open ports. Administrators or attackers use this method to take an inventory of the systems and services on a particular network. These banners often detail what service version is running on the open port.

PASSWORD CRACKERS & COMMAND LINE TOOLS

Password cracking is useful in many circumstances. Whether you are trying to crack a password encrypted file, a WEP network key, or a web application login, there are a variety of public tools available to test the security of any password protected resource. Aircrack-ng is a suite of tools utilized for a variety of things, including breaking into WEP networks by cracking the encryption key. This is accomplished by capturing initialization vectors and brute forcing the key. Medusa is another open source tool used by penetration testers to brute force login credentials. It is commonly used to brute force SSH logins on remote servers. When given a password list, host, and username, it has a good probability of obtaining credentials for a given user. Medusa can also be utilized to brute force web login forms.

Command-line network tools provide a quick and easy way to access network configurations and troubleshoot information. The ping command allows you to check whether a remote system is accessible. Utilizing ICMP (Internet Control Message Protocol), the system that initiates the ping sends an ICMP echo request packet to the target system. This packet requests a response from any device that may be active. The remote system responds to the ping request by sending back an ICMP echo reply. When you receive a successful response to your ping request, you know that network connectivity between the two systems is working properly. This is especially helpful if you're having connectivity issues and aren't sure where the issue resides. For example, if you're trying to access a website, and it's not responding, the issue could be in a variety of places.

Netstat (Network Statistics) can be utilized to view the statistics for network connections on a given device. It may be used to show active connections as well as routing tables. You can even view specific protocols, such as TCP or UDP, by setting certain flags. Traceroute is a command that will map the path a packet takes between two destinations. It also displays transit delays of packets moving across networks. On Linux or MacOS the command is 'traceroute' and on Windows it is 'tracert'. NsLookup and dig (Domain Information Groper) are similar DNS troubleshooting tools. Both show the hostname and IP address of the DNS server that is configured for the local system, and then display a command prompt for further queries. NsLookup is the depreciated version for DNS queries whereas dig is the recommended tool. The arp (Address Resolution Protocol) command allows you to associate a local IP address with a MAC address of local devices based on a cache stored on every computer.

There are a few options when viewing the assigned local IP address configuration of a computer based on the OS and installed network tools. Windows uses 'ipconfig', while Linux and MacOS have either the 'ip' or 'ifconfig' commands available. It will show the hardware address for every network adapter on your machine. Tcpdump is a command line packet analyzer that allows you to capture network traffic and view it in real time with many applicable filters. The Windows equivalent is WinDump. Netcat is a tool that can read or write information to or from network connections using TCP or UDP. Netcat is designed to be a dependable back-end that can be used directly or easily driven by other programs and scripts.

SANITIZATION, STEGANOGRAPHY

In the process of changing hard drives, eliminating any trace of remaining data is ideal. Rather than physically destroying the old drive, overwriting it with new information will permanently destroy any data. Sanitizing the drive can also be accomplished with a tool such as DBAN (Derik's Boot and Nuke). The program is designed to securely erase a hard disk until its data is permanently removed and no longer recoverable, which is achieved by overwriting the data with pseudorandom numbers. Deleting individual files or a set of folders can be achieved with Microsoft SDelete. You can use SDelete both to securely delete existing files, as well as to securely erase any file data that exists in the unallocated portions of a disk including files that you have already deleted or encrypted. Deleting the systems cache and temporary files will finalize deletion of any potentially unwanted duplicate files.

A way to store data in plain sight, but still hide it, is by using steganography, which allows you to embed data inside other non-secret text or data. A common usage of steganography is to embed data inside of an image. Capable software hides the data in an image's pixels to give the illusion that the image is like any other. The image looks identical to the naked eye, but the data can be extracted by steganography software. Steganography can also include concealing data in packets flowing across a network as well as printing microscopic dots on a piece of paper which details the printer model and a timestamp.

EXPLOITATION FRAMEWORKS

Vulnerabilities exist in browsers, operating systems, and applications in use. Rather than writing exploits themselves, attackers will use an exploitation framework with preloaded exploits to attack targets. These frameworks are loaded with exploits current and non-current. It is also possible to load exploits into the framework that do not already exist in it. The mechanisms for delivering the payload and having it execute on the remote system are already built into the framework.

Security professionals engaging in penetration tests need to be able to use the same tools available to attackers seeking to exploit security controls. An exploitation framework offers an efficient way to do that. Metasploit is the most common exploitation framework and is often referred to as a hacker's Swiss army knife. It provides an extensible way to test vulnerabilities using modular plugins. The same flexibility that makes Metasploit an excellent security testing tool also makes it a powerful weapon for attackers. Metasploit began its life as an open source project but was later purchased by the security firm Rapid 7. Because of this heritage, there are now two versions of Metasploit available. The Metasploit community edition remains free, while the Metasploit Pro addition is a commercial product which has extra features.

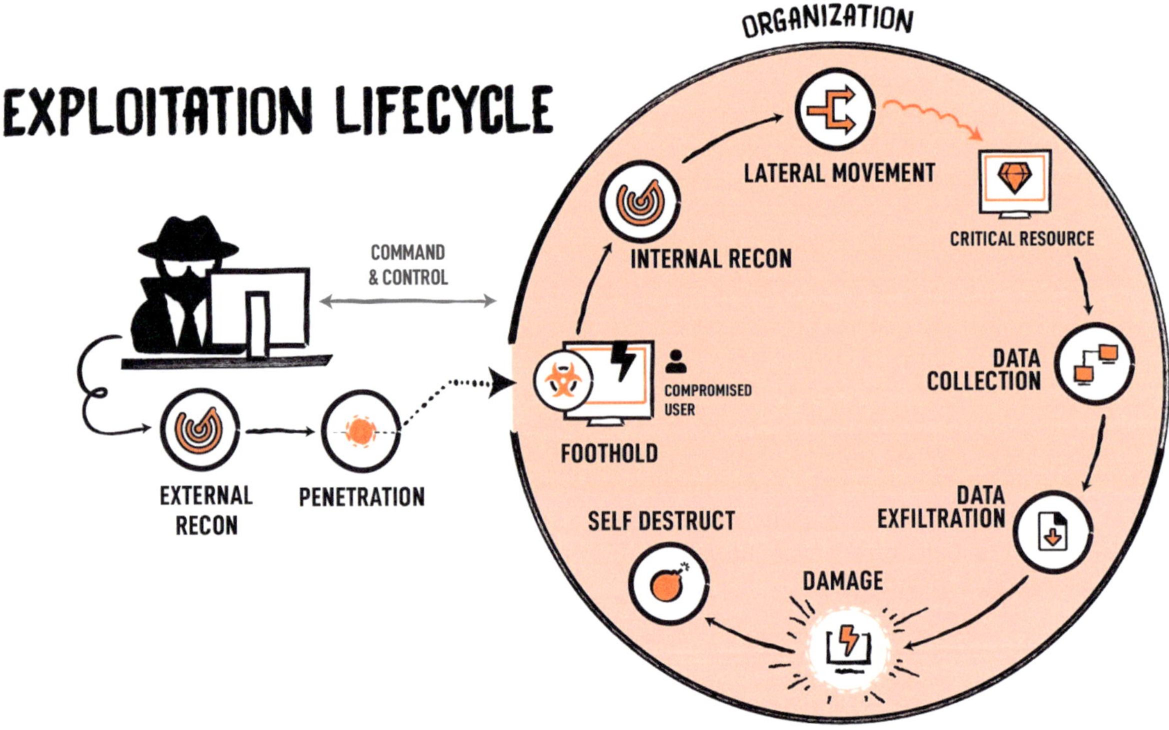

Another common framework is BeEF (Browser Exploitation Framework). As titled, it hosts exploits for all types of internet browsers including Internet Explorer, Firefox, and Google Chrome. BeEF can work in conjunction with other frameworks such as Metasploit to achieve a backdoor shell. RouterSploit, much like BeEF, targets routers specifically. It contains a host of exploits that enable an attacker to bypass administrative login pages and access controls. It may also allow you to place a rootkit on the router which would permit advanced malicious features.

TECHNOLOGIES: CONFIGURATION

SECURITY ISSUES MITIGATION

Unencrypted credentials and clear text passwords are a threat to every organization due to data continually being transported across networks. The intrinsic weakness of some network protocols such as FTP, SMTP, IMAP, Telnet, and HTTP are vulnerable to simple packet capturing attacks. Anything that is sent across the network using any of these protocols is susceptible to being intercepted and read in plain text which could compromise secure credentials.

LOGS: Collecting logs from as many network devices as possible, such as routers, switches, firewalls, IPS, IDS, servers, and application devices, then consolidating them into a SIEM, will give a perspective of what is happening on a network. Extensive reporting, analysis, and event correlation can be done when anomalies occur.

PERMISSION ISSUES: These are many times caused by small oversights. When a server does not have correctly configured directories, they can be accessible to anyone who knows the directory path. On the initial install and build of a server, an audit should take place to verify the correct permissions are applied to all files. Periodic audits should also take place to confirm that no permissions have changed.

ACCESS VIOLATIONS: These may occur when programming errors are made. A segmentation fault is an error that occurs on your OS when an application tries to access an area of memory that it should not have access to. Operating systems constantly check to verify applications are not attempting to break boundaries in place. This is exactly what some malware attempts to do to obtain root or system level privileges on a machine.

CERTIFICATE TRUSTS: What all devices and applications rely on to either identify a device or provide encryption between devices. An internal or external certificate trust authority should sign these certificates as validation of resources accessed. Periodically updating certificates exemplifies best practices. Validation of certificates by applications is crucial to avoid any form of man in the middle attack.

STEALTHILY EXFILTRATING DATA: The goal of any attacker who seeks to obtain sensitive or valuable information. The use of USB drives or CD-ROM to exfiltrate data is common due to the inability to detect it. Commonly, corporations sustain high-speed internet which only enables attackers to exfiltrate data at a high rate if they have obtained access to a system.

ROUTERS, SWITCHES, AND FIREWALLS: These may have a default password that is used to access the control panel. This default password is intended to be changed once an administrator has made changes. However, this is not always the case. Default password attacks make it easy for someone to authenticate to a device and make changes of their own. Misconfigurations, such as outdated exploitable software on a device or debugging information being displayed incorrectly, would allow an attacker to quickly gain access. A potential threat with firewall rules is that they could provide too much access to networked resources. The complexity of organizations with a high volume of rules could make it very difficult to audit.

HUMANS: The weakest link in an organization due to their tendency to make mistakes. If an AUP (Acceptable Use Policy) is broken, personal data may have accidentally or insecurely been transferred. Insider threats already have access to the internal network, which means they are far more capable of executing a malicious plan. Often successful, social engineering techniques take advantage of the nature of people simply wanting to help out. This could lead to compromise of a system or network credentials. Social media could be a large threat to an organization if an employee already has much personal or private information posted for anyone to see. Connecting with a third party email from your work email automatically discloses some information about your company. Emails also imply endorsement by the organization. Company resources should not be utilized for any threatening communication.

UNAUTHORIZED SOFTWARE: Generally prohibited in any organization. This is because the source of software may not be trusted. Malware, spyware, and viruses could be included in the software installed. Third-party software could also cause a conflict with mission-critical software. Software licensing could also come into play if the third-party software was installed illegally. This could cause major repercussions if not handled properly. IT cannot provide ongoing support to an unauthorized piece of software which could cause a security vulnerability if the software is not continually patched.

DOCUMENTATION: This is a vital process allowing any organization to identify when a change occurs anywhere on the network helps to mitigate any threats entering the network or taking place. A baseline of hardware, software, network traffic, and data storage is a way to tell if, when, or what caused a change. Regardless if a piece of malware or a legitimate process altered a system, a change log can be sent to the IT department to determine the next course of action. This process of validation is what takes place before a user authenticates to a secure organization's network via a VPN.

OPERATING SYSTEMS, HARDWARE APPLIANCES, & APPLICATIONS: These all fall under software licensing laws. The reliability of an organization may falter if an application or resource suddenly becomes unavailable due to a license expiring. The integrity of data may rely on a valid license. All data and applications must be accurate and complete.

ASSET MANAGEMENT: Usually an automated process which is constantly updated in a master database. This helps keep control and inventory of where the organization's most valuable assets are. This also helps keep track of the necessary quantity of hardware devices and software licenses you will need network wide.

AUTHENTICATION FACTORS: These include username and password, thumbprint, or one time access code, may lead to security issues. If any step in the authentication of a user fails, this will inhibit them from accessing resources they need. A failure in this process could also grant someone access they were not permitted.

UNDERSTANDING SECURITY TECHNOLOGIES OUTPUT

HOST-BASED IDS: HIDS monitors a local machine for symptoms of unwanted activity. The output of a HIDS or HIPS can include alerts or reports. An alert is an indicator of an immediate event that has just occurred or is continuing to occur. A report is a record of alerts and other information related to the time frame of monitoring and the systems being monitored.

ANTIVIRUS SOFTWARE: An essential security application. The output from an antivirus or antimalware product is an alert or alarm when malware is discovered by the live system monitor, or a report is discovered by the system-wide file scan. A live malware event may cause the antivirus product to respond automatically or prompt the user. Anti-Spam software is a variation on the theme of antivirus software. Spyware monitors your actions and transmits important details to a remote system that spies on your activity. Adware, although quite similar to spyware in form, has a different purpose.

FILE INTEGRITY CHECKING: The activity of comparing the current hash of a file to the stored/previous hash of a file. The output of a file integrity check utility can be analyzed and interpreted in order to understand which files did not have integrity. With this knowledge, it may be possible to review file change logs to determine when the files were modified and what person or software performed the modifications.

HOST-BASED / PERSONAL SOFTWARE FIREWALL: This is a security application that is installed on client systems. A host-based firewall is used to provide protection for the client system from the activities of the user and from communications from the network or Internet. A personal firewall must be kept current with patches and updates. The output of a host-based firewall may be to prompt the user whether or not to grant outbound communication privileges to a software program or alert the user of an attempt to violate existing inbound and outbound firewall rules. If a valid program is requesting network access, it can be granted. But if network access is not authorized or the program is unknown, this request should be denied.

APPLICATION WHITELISTING: A security option that prohibits unauthorized software from executing. Whitelisting is also known as *deny by default* or *implicit deny*. A whitelisting solution may produce an output report detailing the attempts to launch unapproved software with a record of each approved software's execution. This report can be used to determine whether additional software should be added to the whitelist or if any approved applications should be reconsidered.

REMOVABLE MEDIA DRIVES: Removable storage in general is considered both a convenience and a security vulnerability. Any removable media can typically be secured using file-by-file encryption or whole-drive encryption. This may let you move the media from place to place with reasonable assurance that the stored data can't be easily accessed if lost or stolen.

ADVANCED MALWARE TOOLS: These may relate to scanners that include ransomware, rootkits, and potentially unwanted programs, PUPs, in their detection database. When an advanced malware tool detects an unapproved executable, it alerts the administrator. The tool should also initiate an automated removal or quarantine whenever possible. The output of advanced malware tools can be analyzed and interpreted to discover business tasks that are exposing the organization to infection as well as to pinpoint which users are performing risky behaviors.

PATCH MANAGEMENT: The formal process of ensuring that updates and patches are properly tested and applied to production systems. The output from a patch management tool will be a report indicating that monitored systems are or are not in compliance with the approved updates. Some patch management systems may be able to automatically apply approved patches, whereas others require the administrator to install updates manually.

ALL-IN-ONE SECURITY APPLIANCE: A hardware device designed to operate inline between an Internet connection and a network. Its goal is to detect and filter all manners of malicious, wasteful, or otherwise unwanted traffic. These devices can be called security gateways or unified threat management, UTM, systems. The output of the UTM will either be unique reports from each of the sub-features or a single report with an amalgamation of results from all tools. Responses to the UTM report should be based on the specific item discovered and will follow the same procedures as discussed earlier in this section.

DATA LOSS PREVENTION: DLP is a system designed to reduce and/or prevent data loss or data leakage to external unauthorized entities. If a violation of DLP occurs, its report should indicate the data that was involved, the user(s) related to the breach, and the applications involved in the exfiltration. The response to a DLP issue is to evaluate the value of the leaked asset to determine the severity or priority of the response. It may dictate that legal proceedings be started against the internal and external entities involved with the violation. The user involved with violating DLP restrictions may need to be retrained, have their job privileges reduced, and potentially be terminated. The applications and systems that enabled the loss event to occur need to be adjusted to block the reoccurrence of the violation.

DATA EXECUTION PREVENTION: DEP is a memory security feature of many operating systems aimed at blocking a range of memory abuse attacks, including buffer overflows. When DEP fails, there may not be any specific official output, log, or report of the situation. Some DEP solutions may create a memory abuse attempt log, but it may not record events that were designed to specifically violate the DEP protections. Often the only result is when an administrator happens to notice that malware or unauthorized code is executing on a system where DEP was present. The response should be to terminate the offending code and remove it from the system.

WEB APPLICATION FIREWALL: A device, server add-on, virtual service, or system filter that defines a strict set of communication rules for a website and all visitors. The output of a web application firewall or web security gateway is similar to that of a firewall, IDS, or UTM. The output may be a real-time alert or an after-the-fact report. The output should be evaluated and appropriate responses determined.

SECURE MOBILE DEVICE DEPLOYMENT

CONNECTION METHODS: A cellular network or a wireless network is the primary communications technology used by many mobile devices, especially cell phones and smartphones. The network is organized around areas of land called cells, which are centered on a primary transceiver, known as a cell site, cell tower, or base station. Generally, cellular service is encrypted, but only while the communication is being transmitted from the mobile device to a transmission tower. Communications are effectively plain text once they are being transmitted over wires. So, avoid performing any task over cellular that is sensitive or confidential in nature. Use an encrypted communications application to pre-encrypt communications before transmitting them over a cellular connection.

- **WIFI:** Wireless networking was originally defined by the IEEE 802.11 standard. WiFi is a nearly ubiquitous communication scheme available in most homes, offices, and public retail locations, such as restaurants and stores.
- **SATCOM:** Satellite communication is a means of audio and data transmission using satellites orbiting in near-earth orbit. SATCOM devices benefit from nearly complete service coverage, thanks to the broad footprint of a signal transmitted from 100+ miles above the surface of the planet. The data transmission speeds of SATCOM are rather poor compared to those of terrestrial solutions, but it may be the only available option in remote locations.
- **BLUETOOTH:** Plain text by default in most implementation and usage scenarios, but can be encrypted with specialty transmitters and peripherals. Bluetooth operates between devices that have been paired, which is a means of loosely associating devices with each other using a default pair code. Since Bluetooth is typically a plain-text communication, do not use it to support sensitive or confidential transactions. Use an alternate means of communications that can provide encrypted transactions. Even if you are using a special implementation of Bluetooth that does encrypt the wireless signal, that encryption ends at the Bluetooth transmitter/receiver device on each end of the wireless signal.
- **NEAR FIELD COMMUNICATION:** NFC is a standard to establish radio communications between devices in close proximity. NFC is designed to be a secure communications system, and its signals are encrypted or encoded in most cases. NFC is not used to support ongoing or large data transmissions, such as WiFi, cellular, or even Bluetooth, so the risks are minimal simply based on its limited data transmission uses.
- **ANT:** A proprietary protocol owned by Garmin with open access multicast sensor network technology. ANT offers the ability to encrypt communications, but it is not always enabled. Some implementations of ANT, such as ANT+, do not offer any encryption because they focus on cross-vendor interoperability rather than security. Similar to NFC, ANT has limited risk due to its current use limitations. However, always be cautious when using any plain text communications system.

- **INFRARED**: Not as common a communication technology as wireless is for modern devices. However, there are still plenty of infrared implementations. It is unlikely you will use infrared communications; if you do, however, be cautious of transmitting valuable or sensitive data. Some modern mobile phones continue to include an infrared port for use as a transmitter for controlling televisions and other A/V entertainment equipment.
- **UNIVERSAL SERIAL BUS**: USB is a standard for connecting peripheral devices and primary computers over a wired link. Once devices are connected via USB, they typically appear in standard file management tools as USB storage devices, where reading and writing of data can take place. The only real protection provided by USB is that it is a wired connection, as opposed to wireless, and that an encrypted and screen-locked device is likely to disable the USB port. Only when the screen lock is cleared does the USB port become enabled for data exchange.

MOBILE DEVICE MANAGEMENT CONCEPTS: Application control or application management is a device management solution that limits which applications can be installed onto a device. Although security features must be enabled to have any beneficial effect, it's just as important to remove apps and disable features that aren't essential to business tasks or common personal use. The wider the range of enabled features and installed apps, the greater the chance that an exploitation or software flaw will cause harm to the device and/or the data it contains.

- **CONTENT MANAGEMENT**: is the control of mobile devices and their access to content hosted on company systems as well as controlling access to company data stored on mobile devices. Typically an MCM, mobile content management system, is used to control company resources and the means by which they are accessed or used on mobile devices. The goal of a content management system for mobile devices is to maximize performance and work benefit while reducing complexity, confusion, and inconvenience. An MCM may also be tied to an MDM to ensure secure use of company data.
- **REMOTE WIPE**: Lets you delete all data and possibly even configuration settings from a device remotely. However, a remote wipe isn't a guarantee of data security. A skilled thief may be able to undelete data files after the wiping process. A way to improve the benefit of remote wipe is to keep the mobile device's storage encrypted. Thus, an undelete operation would only recover encrypted files and most likely not allow the attacker to decode the data.
- **GEO-FENCING**: The designation of a specific geographical area that is then used to implement features on mobile devices. A geo-fence can be defined by GPS coordinates, wireless indoor positioning system, IPS, or presence or lack of a specific wireless signal. A device can be configured to enable or disable features based on a geo-fenced area.
- **GEO-TAGGING**: Also called geo-location; this is the ability of a mobile device to include details about its location in any media created by the device. Geolocation data is commonly used in navigation tools and by many location-based services. Once a geotagging photo has been uploaded to the Internet, a potential cyber-stalker may have access to more information than the uploader intended. This is prime material for security-awareness briefs for end users.

- **A SCREEN LOCK**: Designed to prevent someone from casually picking up and being able to use your phone or mobile device. However, most screen locks can be unlocked by swiping a pattern or typing a number on a keypad display. The lockout feature ensures that if you leave your device unattended or it's lost or stolen, it will be difficult for anyone else to be able to access your data or applications.
- **PUSH NOTIFICATIONS**: Send information to your device rather than have the device pull information from an online resource. Many apps and services can be configured to use push and/or pull notifications.
- **STRONG PASSWORD**: A great idea on a phone or other mobile device if locking the phone provided true security. But most mobile devices aren't secure, so even with a strong password, the device is still accessible over Bluetooth, wireless, or a USB cable. As mentioned previously, it's also prudent to combine device authentication with device encryption to block access to stored information via a connected cable.
- **BIOMETRICS**: A convenient means of authenticating to mobile devices. However, they are not as accurate as we may wish them to be. A biometric only has to satisfy an approximation of the reference profile of the stored biometric value. Thus, biometrics should not be employed as the only means or mechanism to authenticate to a device. If the device holds highly valuable and sensitive content, then don't use single-factor biometrics. Instead use a biometric only as one element of a multifactor authentication.
- **CONTEXT-AWARE AUTHENTICATION**: An improvement on traditional authentication means. Contextual authentication evaluates the origin and context of a user's attempt to access a system. If the user originates from a known trusted system, such as a system inside the company facility, then a low-risk context is present and a modest level of authentication is mandated for gaining access. If the context and origin of the user is from an unknown device and/or external/unknown location, the context is high risk.
- **CONTAINERIZATION**: The next stage in the evolution of the virtualization trend for both internally hosted systems and cloud providers and services. Containerization is based on the concept of eliminating the duplication of OS elements and removing the hypervisor altogether. Instead, each application is placed into a container that includes only the actual resources needed to support the enclosed application. Containerization can be used in relation to mobile devices by hosting the primary OS on a containerization host in the company cloud so that the actual mobile device is only used as a remote control interface to the OS container rather than having business apps and company data on the device itself.
- **STORAGE SEGMENTATION**: Used to artificially compartmentalize various types or values of data on a storage medium. On a mobile device, the device manufacturer and/or the service provider may use storage segmentation to isolate the device's OS and preinstalled apps from user-installed apps and user data. This allows for ownership and rights over user data to be retained by the user, while granting ownership and rights over business data to the organization, even on devices owned by the employee.
- **ENCRYPTION**: Often a useful protection mechanism against unauthorized access to data, whether in storage or in transit. Most mobile devices provide some form of storage encryption. When this is available, it should be enabled.

Enforcement and monitoring of most other sources of apps for either smart device platform are labeled as third-party app stores. These app stores often have less rigorous rules regarding hosting an app. When a mobile device is being managed by an organization, especially when using an MDM, most third-party sources of apps will be blocked. Such third-party app sources represent a significant increase in risk of data leakage or malware intrusion to an organizational network.

- **ROOTING / JAILBREAKING:** The action of breaking the digital rights management (DRM) security on the bootloader of a mobile device in order to be able to operate the device with root or full system privileges. An organization should prohibit the use of rooted devices on the company network or even access to company resources whenever possible. For some, a rooted device may provide benefits that exceed the risks, but such devices should be operated as stand-alone equipment and not as endpoint devices of a company network.
- **SIDELOADING:** The activity of installing an app onto a device by bringing the installer file to the device through some form of file transfer or USB storage method rather than installing from an app store. Most organizations should prohibit user sideloading, because it may be a means to bypass security restrictions imposed by an app store or the MDM.
- **FIRMWARE:** Mobile devices come preinstalled with a vendor, telco-provided firmware, or core operating system. The firmware can be updated by an upgrade provided by the vendor or telco. An organization should not allow users to operate mobile devices that have custom firmware unless that firmware is preapproved by the organization. Some custom firmware can be returned to a non-rooted state once the firmware install is completed.
- **CARRIER LOCK:** Most mobile devices purchased directly from a telco are carrier locked. This means you are unable to use the device on any other telco network until the carrier lock is removed or carrier unlocked. A carrier unlocked device should not represent any additional risk to an organization; thus there is likely no need for a prohibition of carrier unlocked devices on company networks.
- **FIRMWARE OTA UPDATES:** Upgrades, patches, and improvements to the existing firmware of a mobile device that are downloaded from the telco or vendor over the air. Organizations should attempt to test new updates before allowing managed devices to receive them. There simply may need to be a waiting period established so the MDM vendor can update their management product to properly oversee the deployment and configuration of the new firmware update.
- **CAMERAS:** The company security policy needs to address mobile devices with onboard cameras. Some environments disallow cameras of any type. If geo-fencing is available, it may be possible to use MDM to implement a location-specific hardware-disable profile in order to turn off the camera while the device is on company premises but return the feature to operational status once the device leaves the geo-fenced area.

- **SHORT MESSAGING SERVICE**: SMS, also known as texting, and MMS, Multimedia Messaging Service, are communication functions provided by telcos, most commonly used on mobile devices. SMS and MMS represent generally the same level of risk and benefit as that of email. It's a good idea to block attachments and file exchange. Spam filtering is needed, though it may be called SPIM (Spam over Instant Message). Social engineering defenses must be established. Users must be trained on avoiding risk and minimizing distractions.
- **REMOVABLE STORAGE**: Some smartphones support microSD cards, whereas most larger mobile devices, such as tablets and notebook computers, support SD cards and other media card formats, which can be used to expand available storage on a mobile device. Organizations need to consider whether the use of removable storage on portable and mobile devices is a convenient benefit or a significant risk vector. If the former, proper access limitations and use training are necessary. If the latter, then a prohibition of removable storage can be implemented via MDM.
- **USB SPECIFICATIONS**: Can allow a mobile device with a USB port to act as a host and use other standard peripheral USB equipment, such as storage devices, mice, keyboards, and digital cameras. USB OTG is a feature that can be disabled via MDM if it is perceived as a risk vector for mobile devices used within an organization.
- **SPEAKER**: Most mobile devices with a speaker also have a microphone. The microphone can be used to record audio, noise, and voices nearby. If microphone recording is deemed a security risk, this feature should be disabled using an MDM or deny presence of mobile devices in sensitive areas or meetings.
- **GPS TAGGING**: The same as geolocation and geotagging. This is also a feature that can be disabled using an MDM.
- **WIFI DIRECT**: The new name for the wireless topology of ad hoc or peer-to-peer connections. In a business environment, WiFi Direct should only be used where WPA-2 can be used. Otherwise, the plain-text communication presents too much risk.
- **TETHERING**: The activity of sharing the cellular network data connection of a mobile device with other devices. The sharing of data connection can take place over WiFi, Bluetooth, or USB cable. Tethering may represent a risk to the organization. It is a means for a user to grant Internet access to devices that are otherwise network isolated, and it can be used as a means to bypass the company's filtering, blocking, and monitoring of Internet use. Thus, tethering should be blocked while a mobile device is within a company facility.
- **MOBILE DEVICE–BASED PAYMENT SYSTEMS**: An organization is unlikely to see any additional risk based on mobile payment solutions. However, caution should still be taken when implementing them on company-owned equipment or when they are linked to the company's financial accounts.

DEPLOYMENT MODELS

BYOD: A policy that allows employees to bring their own personal mobile devices to work and use those devices to connect to business resources and/or the Internet through the company network. Users need to understand the benefits, restrictions, and consequences of using their own devices at work. Reading and signing off on the BYOD policy, along with attending an overview or training program, may be sufficient to accomplish reasonable awareness.

CORPORATE OWNED, PERSONALLY ENABLED: This concept, COPE, means the organization purchases devices and provides them to employees. Each user is then able to customize the device and use it for both work activities and personal activities. COPE allows the organization to select exactly which devices are to be allowed on the organizational network—specifically only those devices that can be configured into compliance with the security policy.

CHOOSE YOUR OWN DEVICE: The concept of CYOD provides users with a list of approved devices from which to select the device to implement. A CYOD can be implemented so that employees purchase their own devices from the approved list or the company can purchase the devices for the employees.

CORPORATE-OWNED MOBILE STRATEGY: When the company purchases mobile devices that can support compliance with the security policy. These devices are to be used exclusively for company purposes, and users should not perform any personal tasks on them. This often requires workers to carry a second device for personal use.

VIRTUAL DESKTOP INFRASTRUCTURE: VDI is a means to reduce the security risk and performance requirements of end devices by hosting virtual machines on central servers that are remotely accessed by users. This has led to virtual mobile infrastructure, VMI, where the operating system of a mobile device is virtualized on a central server. Thus most of the actions and activities of the traditional mobile device are no longer occurring on the mobile device itself. This remote virtualization allows an organization greater control and security than when using a standard mobile device platform. It can also enable personally owned devices to interact with the VDI without increasing the risk profile. This concept requires a dedicated, isolated wireless network to keep BYOD devices from interacting directly with company resources other than through the VDI solution.

SECURITY PROTOCOLS IMPLEMENTATION

DOMAIN NAME SYSTEM: DNS is the hierarchical naming scheme used in both public and private networks. DNS links IP addresses and human-friendly fully qualified domain names (FQDNs) together. A FQDN consists of three main parts:

- Top-level domain (TLD): The com in www.google.com
- Registered domain name: The google in www.google.com
- Subdomain(s) or hostname: The www in www.google.com

DOMAIN NAME SYSTEM SECURITY EXTENSIONS: DNSSEC is a security improvement to the existing DNS infrastructure. The primary function of DNSSEC is to provide reliable authentication between devices during DNS operations. DNSSEC has been implemented across a significant portion of the DNS system. Each DNS server is issued a digital certificate, which is then used to perform mutual certificate authentication. The goal of DNSSEC is to prevent a range of DNS abuses where false data can be injected into the resolution process. Once fully implemented, DNSSEC will significantly reduce server-focused DNS abuses.

SECURE SHELL: SSH is a secure replacement for Telnet and many of the Unix "r" tools, such as rlogon, rsh, rexec, and rcp. SSH is the protocol most frequently used with a terminal editor program such as HyperTerminal in Windows, Minicom in Linux, or PuTTY in both. An example of SSH use would involve remotely connecting to a switch or router in order to make configuration changes. SSH transmits both authentication traffic and data in a secured encrypted form. Thus, no information is exchanged in clear text. SSH is a very flexible tool. It can be used as a secure Telnet replacement; it can be used to encrypt protocols similar to TLS, such as SFTP; and it can be used as a VPN protocol.

S/MIME: An Internet standard for encrypting and digitally signing email; it takes the standard MIME element of email, which enables email to carry attachments and higher-order textual information, and expands this to include message encryption.

The basic process is as follows:

- The sender's system generates a random symmetric key.
- The sender encrypts the message with the random symmetric key.
- The symmetric key is enveloped (encrypted) using the recipient's public key.
- The message and the envelope are sent to the recipient.
- The recipient opens (decrypts) the envelope using the recipient's private key to extract the symmetric key.
- The recipient decrypts the message using the symmetric key.

SECURE REAL-TIME TRANSPORT PROTOCOL: Secure RTP or SRTP is a security improvement on RTP (Real-Time Transport Protocol) used in many VoIP (Voice over IP) communications. SRTP aims to minimize the risk of VoIP DoS through robust encryption and reliable authentication.

LIGHTWEIGHT DIRECTORY ACCESS PROTOCOL: LDAP is a standardized protocol that enables clients to access resources within a directory service. A directory service is a network service that provides access to a central database of information, which contains detailed information about the resources available on a network. It's important to secure LDAP rather than allow it to operate in a plain-text insecure form. This is accomplished by enabling the Simple Authentication and Security Layer, SASL, on LDAP, which implements Transport Layer Security, TLS, on the authentication of clients as well as all data exchanges. This results in LDAP Secured, LDAPS.

FTP SECURE: FTPS (or FTP SSL, a variation of FTP secured by SSL, now TLS) is an FTP service variation distinct from SSH-secured FTP (SFTP). FTPS is supported by FTP servers in either an implicit or an explicit mode, FTPIS or FTPES. Implicit implies that the client must specifically challenge the FTPS server with a TLS/SSL ClientHello message. This assumes that only FTPS clients will connect. In order to allow traditional FTP clients to continue to operate over ports 20, data channel, and 21, control channel, FTPS is delegated to ports 990, control channel, and 989, data channel. It's important, however, to note that implicit mode is now considered deprecated. Explicit, FTPES, mode implies that the FTPS client must specifically request an FTPS connection on ports 20 and 21; otherwise, an insecure FTP connection will be attempted. More information regarding explicit mode is available in RFC 2228 and RFC 4218.

SECURE FTP: An encrypted alternative to standard FTP. It encrypts both authentication and data traffic between the client and server by employing SSH to provide secure FTP communications. Thus, SFTP provides protection for both the authentication traffic and the data transfer occurring between a client and server. No matter what secure FTP solution is employed, both the server and the client must have the same solution. The client and the server must have compatible or interoperable FTP tools to establish a connection and support the exchange of files. Otherwise, FTP session establishment and subsequent file-transfer communications won't be possible.

SIMPLE NETWORK MANAGEMENT PROTOCOL: SNMP is a standard network-management protocol supported by most network devices and TCP/IP-compliant hosts. SNMP operates over UDP ports 161 and 162. UDP port 161 is used by the SNMP agent, that is, network device, to receive requests, and UDP port 162 is used by the management console to receive responses and notifications, also known as trap messages.

: SSL and TLS are used to encrypt traffic between a client and server. Through the use of SSL or TLS, web surfers can make online purchases, interact with banks, and access private information without disclosing the contents of their communications. SSL and TLS can make web transactions private and secure.

When you use SSL / TLS to secure communications, a multistep handshake process must be completed to establish the secured session:

- The client requests a secure connection.
- The server responds with its certificate, the name of its certificate authority, issuing CA, and its public key.
- The client verifies the server's certificate, produces a session (symmetric) encryption key, encrypts the key with the server's public key, and sends the encrypted key to the server.
- The server unpacks the session key and sends a summary of the session details to the client, encrypted with the session key.
- The client reviews the summary and sends its own summary back to the server, likewise encrypted with the session key.
- After both entities receive a matching session summary, secured SSL communications are initiated.

SSL / TLS uses symmetric keys as the session keys. The session keys available for SSL include 40-bit and 128-bit strengths. TLS session keys can currently span between 128 bit and 256 bit.

HYPERTEXT TRANSFER PROTOCOL: HTTP operates over TCP port 80. It's a plain-text or clear-text communication protocol; thus, it offers no security or privacy to transactions. When SSL or TLS is used to secure transactions, this is known as Hypertext Transfer Protocol over SSL, HTTPS, or Hypertext Transfer Protocol Secured, HTTPS. You can recognize when secure web communications are occurring using SSL or TLS because the URL begins with HTTPS and a locked padlock icon appears in the status bar at the bottom of the browser window.

POST OFFICE PROTOCOL: POP and IMAP (Internet Message Access Protocol) are secured by implementing SSL / TLS, encryption. POP and IMAP are email retrieval protocols, unlike SMTP which is an email sending or forwarding protocol. When using SMTP, POP, or IMAP in their encrypted form, the security is only between the email client and the local email server. Any subsequent email communications between email servers may or may not be using encrypted SMTP. But those communication pathways are not under the control of the email client system, but of the administrators of each SMTP server.

Voice communications have shifted from traditional landlines to mobile phones employing cellular to VoIP. A few VoIP solutions can provide end-to-end encryption. This does not seem to be the industry standard, and often encryption is not possible between dissimilar systems. Whenever you have the option to employ an end-to-end encryption VoIP system, it will be a better security choice than any other.

TIME SYNCHRONIZATION: An important element of security management as well as overall network and system management. Many essential functions and services are dependent on reliable time information. The protocol Network Time Protocol, NTP, which operates over UDP port 123, is used to synchronize system clocks with each other and with an external reliable time source. While some cryptography is available in NTPv4, it is mostly for integrity checking and not for confidentiality. It may be necessary to implement IPSec or other VPN sessions between systems and then tunnel the NTP through the encrypted channel.

EMAIL & WEB COMMUNICATION PROTECTION: Can be achieved easily using TLS-encrypted forms of their respective protocols. See "S/MIME," "SSL/TLS," "HTTPS," and "Secure POP/IMAP." Secure file transfer is easily accomplished using either SSH or TLS-encrypted forms of FTP; see the sections "SSH," "FTPS," "SFTP," and "SSL/TLS". It is fairly common to use the native Windows service of Server Message Block (SMB) or the Network File System (NFS) on Linux and Unix. These file transfer tools are convenient but they are plain text. You should establish an IPSec or other VPN connection, and then encapsulate the file transfer session within the VPN in order to gain security for these tools.

DIRECTORY SERVICE: A key feature of modern networks; most clients should support the secured form of directory service interaction, such as LDAPS (or SASL), so that secure means are required.

REMOTE ACCESS SERVER: RAS is a network server supporting connections from distant users or systems. RAS often supports modem banks, VPN links, and even terminal services connections.

MODEM: A device that creates a network communication link between two computers / networks over a telephone line. Modems are one of the slowest remote-connection methods still widely supported by OSs. A common security protection added to dial-up modems is callback, a feature that disconnects the remote user immediately after authentication and then calls back the remote user at a predefined number. Callback ensures that the authenticated user is located at the correct phone number before access to the network is granted.

REMOTE AUTHENTICATION: A catchphrase that refers to any mechanism used to verify the identity of remote users. Several well-known examples of remote authentication include RADIUS, TACACS, 802.1x, and Challenge Handshake Authentication Protocol, CHAP.

VOICE OVER IP: VoIP is a tunneling mechanism used to transport voice and/or data over a TCP/IP network. VoIP has the potential to replace or supplant PSTN because it's often less expensive and offers a wider variety of options and features. VoIP can be used as a direct telephone replacement on computer networks as well as mobile devices. However, VoIP is able to support video and data transmission to allow videoconferencing and remote collaboration on projects.

DNS SECURITY: This should be taken seriously. If your DNS resolutions are being falsified or monitored, your system and/or organization can be harmed. DNSSEC is a means to improve the security of DNS resolutions.

COMMUNICATIONS BETWEEN ROUTERS & SWITCHES: Yet another potential target for attackers. If they are able to interfere with the convergence of routing tables or STP, then attackers can either redirect traffic down pathways to which they have physical or logical access, or they can simply implement a DoS. Employing VPNs or other encryption services between network management devices can reduce these risks.

SUBNETTING: A divisioning process used on networks to divide larger groups of hosts into smaller collections. Subnetting is mainly a logical activity, but it can be used to direct or guide physical divisioning. In fact, many large organizations mimic their logical subnetting infrastructure in their physical deployment for easier troubleshooting and maintenance. Ultimately, in the TCP/IPv4 protocol, subnetting is defined by the assigned host IP address and its related subnet mask. The subnet mask is a 32-bit binary number that indicates which portions of a host IP address (also a 32-bit binary number, at least for TCP/IPv4) define the network ID (or subnet ID) and which portions define the host ID. Network and subnet IDs are unique within each organization's private network or across the public Internet. Host IDs are unique only within the local subnet. In much the same way that an area code defines the general area where a phone number resides, a network ID defines where a subnet resides. Within one area code and another, there are duplicate seven-digit phone numbers, and within multiple subnets there are duplicate host IDs. However, unlike phone numbers, IP addresses are always presented with their entire complement of numbers and, when necessary or important, their related subnet mask.

SUBSCRIPTION SERVICES: These are becoming a common tool employed by businesses and individuals alike. Subscriptions for all types of services are gaining widespread support; these include email, document editing, cloud storage, cloud backup, gaming, video entertainment, VoIP, and remote hosting. No matter what subscription service is in use, care should be taken to ensure that the connection between the online service server and the client/subscriber/customer system is encrypted. It is also important to determine whether the online service provides proper security for the customer database and any files or data stored online.

ARCHITECTURE & DESIGN

BEST PRACTICES & STANDARDS

INDUSTRY-STANDARD FRAMEWORKS & REFERENCE ARCHITECTURES

SECURITY FRAMEWORK: A guide or plan for keeping your organizational assets safe. It provides a structure to the implementation of security for both new organizations and those with a long history. A security framework should provide perspective that security is not just an IT concern, but an important business operational function.

REGULATORY SECURITY FRAMEWORK: A security guidance established by a government regulation or law. Regulatory frameworks are thus crafted or sponsored by government agencies. Many regulatory frameworks are publicly available and thus can be adopted and applied to private organizations as well. Each organization is unique and thus should use several frameworks to assemble a solution that addresses their specific security needs.

NON-REGULATORY SECURITY FRAMEWORK: Any security guidance crafted by a non-government entity, including open-source communities and commercial entities. Non-regulatory frameworks may require a licensing fee or subscription in order to view and access details of the framework.

NATIONAL SECURITY FRAMEWORK: Any security guidance designed specifically for use within a particular country. The author of a national framework may attempt to restrict access to the details of their framework in order to control or limit implementation to just their local industries. International security frameworks are designed on purpose to be nation independent. These are crafted with the goal of avoiding any country-specific limitations or idiosyncrasies in order to support worldwide adoption of the framework. Compliance with international security frameworks simplifies interactions between organizations located across national borders by ensuring they have compatible and equivalent security protections.

INDUSTRY-SPECIFIC FRAMEWORKS: Are those crafted to be applicable to one specific industry, such as banking, health care, insurance, energy management, transportation, or retail. These types of frameworks are tuned to address the most common issues within an industry and may not be as easily applicable to organizations outside of that target.

BENCHMARKS / SECURE CONFIGURATION GUIDES

BENCHMARK: A documented list of requirements used to determine whether or not a system, device, or software solution is allowed to operate within a securely managed environment.

PLATFORM / VENDOR-SPECIFIC GUIDES: Often quite specific to an operating system/platform, application, or product vendor. These types of guides can be quite helpful in securing a product since they may provide step-by-step, click-by-click, command-by-command instructions on securing a specific application, OS, or hardware product.

Benchmarks and security configuration guides can focus on specific web server products, such as Microsoft's Internet Information Service (IIS) or Apache Web Server.

Benchmarks and security configuration guides can focus on specific operating systems, such as Microsoft Windows, Apple Macintosh, Linux, or Unix.

Benchmarks and security configuration guides can focus on specific application servers, such as Domain Name System, DNS, Dynamic Host Configuration Protocol, DHCP, databases, Network Area Storage, NAS, Storage Area Network, SAN,, directory services, virtual private network, VPN, or Voice over Internet Protocol, VoIP.

Both configurations and benchmarks alike can focus on key network infrastructure devices, such as firewalls, switches, routers, wireless access points, VPN concentrators, web security gateways, virtual machines/hypervisors, or proxies.

GENERAL-PURPOSE SECURITY CONFIGURATION GUIDES: These are more generic in their recommendations rather than being focused on a single software or hardware product. This makes them useful in a wide range of situations, but they provide less detail and instruction on exactly how to accomplish the recommendations.

DEFENSE IN DEPTH (DID) / LAYERED SECURITY

Defense in Depth, or DiD, is the use of multiple types of access controls concentric circles or layers. This form of layered security helps an organization avoid a monolithic security stance. A monolithic mentality is the belief that a single security mechanism is all that is required to provide sufficient security.

VENDOR DIVERSITY: Is important for establishing defense in depth in order to avoid security vulnerabilities due to one vendor's design, architecture, and philosophy of security. No one vendor can provide a complete end-to-end security solution that protects against all known and unknown exploitations and intrusions. Thus, to improve the security stance of an organization, it is important to integrate security mechanisms from a variety of vendors, manufacturers, and programmers.

CONTROL DIVERSITY: This is an essential in order to avoid a monolithic security structure. Do not depend on a single form or type of security; instead, integrate a variety of security mechanisms into the layers of defense. Using three firewalls is not as secure as using a firewall, an IDS, and strong authentication.

ADMINISTRATIVE CONTROLS: These typically include security policies as well as mechanisms for managing people and overseeing business processes. It is important to ensure a diversity of administrative controls rather than relying on a single layer or single type of security mechanism.

TECHNICAL CONTROLS: These include any logical or technical mechanism used to provide security to an IT infrastructure. Diverse and multilayered defenses require a more complex attack approach using numerous exploitations in series successfully and without detection in order to compromise the target.

USER TRAINING: A key part of any security endeavor. Users need to be trained in how to perform their work tasks in accordance with the limitations and restrictions of the security infrastructure. Users need to understand, believe in, and support the security efforts of the organization; otherwise, users will find ways around security controls in order to do their jobs and may cause accidental or intentional security control sabotage.

SECURE NETWORK ARCHITECTURE CONCEPTS

ZONES / TOPOLOGIES

NETWORK ZONE: An area of a network designed for a specific purpose, such as internal use or external use. Network zones are logical and/or physical divisions or segments of a LAN that allow for supplementary layers of security and control. Each security zone is an area of a network that has a single defined level of security. That security may focus on encoding authorized access, preventing access, protecting confidentiality and integrity, or limiting traffic flow.

A DEMILITARIZED ZONE: DMZ is a special-purpose subnet that is designed specifically for low-trust users to access specific systems, such as the public accessing a web server. If the DMZ is compromised, the private LAN isn't necessarily affected or compromised. A DMZ can also be deployed through the use of a multi-homed firewall. Such a firewall has three interfaces: one to the Internet, one to the private LAN, and one to the DMZ

EXTRANET: A privately controlled network segment or subnet that functions as a DMZ for business-to-business transactions. Extranets are based on TCP/IP and often use the common Internet information services, such as web browsing, FTP, and email. Extranets aren't accessible to the general public. They often require outside entities to connect using a VPN.

INTRANET: A private network or private LAN. This term was coined in the 1990s when there was a distinction between traditional LANs and those adopting Internet technologies, such as the TCP/IP protocol, web services, and email. Now that most networks use these technologies, the term intranet is no longer distinct from LAN. All organizations that have a network have an intranet. Thus, any scenario involving a private LAN is also an intranet.

WIRELESS NETWORK: Uses radio waves as the communication media instead of copper or fiber-optic cables. A wireless network zone can be isolated using encryption (such as WPA-2) and unique authentication (so that only users and devices authorized for a specific network zone are able to log into that wireless zone). Wireless is a viable option in workspaces where portable devices are needed, or when running network cables is cost prohibitive.

GUEST ZONE: Area of a private network designated for use by temporary authorized visitors. It allows nonemployee entities to partially interact with your private network, or at least with a subset of strictly controlled resources, without exposing your internal network to unauthorized user threats.

HONEYNET: Consists of two or more networked honeypots used in tandem to monitor or re-create larger, more diverse network arrangements. Often, these honeynets facilitate IDS deployment for the purposes of detecting and catching both internal and external attackers.

NETWORK ADDRESS TRANSLATION: NAT is a common solution for enabling outbound Internet connection, while still protecting systems from direct access by external entities, both benign and malicious. Because systems must have an Internet-capable TCP/IP address in order to communicate across the Internet, and because leasing public IP addresses for every single system on a network is expensive, NAT is often the most cost effective and secure strategy for organizations.

AD HOC: A form of wireless network in which individual hosts connect directly to each other rather than going through a middleman. It is often referred to as a "peer to peer network" as well.

SEGREGATION / SEGMENTATION / ISOLATION

NETWORK SEGMENTATION: Involves controlling traffic among networked devices. Complete or physical network segmentation occurs when a network is isolated from outside communications, so transactions can only occur between devices within the segmented network. Security layers exist where devices with different levels of classification or sensitivity are grouped together and isolated from other groups with different security levels. This isolation can be absolute or one-directional. All networks have scenarios in which segregation, segmentation, and isolation are needed. Without establishing a distinction between internal private networks and external public networks, maintaining privacy, security, and control is very challenging for the protection of sensitive data and systems.

PHYSICAL SEGMENTATION: Occurs when no links are established between networks. This is also known as an air gap. If there are no cables and no wireless connections between two networks, then a physical network segregation/segmentation/isolation has been achieved. This is the most reliable means of prohibiting unwanted transfer of data. However, this configuration is also the most inconvenient for the rare events where communications are desired or necessary.

VIRTUAL LOCAL AREA NETWORK: VLAN is a hardware-imposed network segmentation created by switches. VLANs are used for traffic management. Communications between ports within the same VLAN occur without hindrance, but communications between VLANs require a routing function, which can be provided either by an external router or by the switch's internal software.

VIRTUALIZATION TECHNOLOGY: Used to host one or more OSs in the memory of a single host computer. This mechanism allows virtually any OS to operate on any hardware. It also lets multiple OSs work simultaneously on the same hardware. Virtualized servers and services are indistinguishable from traditional servers and services from a user's perspective. Additionally, recovery from damaged, crashed, or corrupted virtual systems is often quick.

TUNNELING / VPN

VIRTUAL PRIVATE NETWORK: VPN is a communication tunnel between two entities across an intermediary network.

SITE-TO-SITE VPN: A connection between two organizational networks.

REMOTE-ACCESS VPN: This is a variant of the site-to-site VPN. The difference is that one endpoint of a remote-access VPN is the single entity of a remote user that connects into an organizational network.

SECURITY DEVICE / TECHNOLOGY PLACEMENT

SENSOR: Hardware or software tool used to monitor an activity or event in order to record information or at least take notice of an occurrence. For sensors to be effective, they need to be located in proper proximity to the item(s) being monitored. This might require the sensor to monitor all network traffic, monitor a specific doorway, or monitor a single computer system.

SECURITY COLLECTOR: Any system that gathers data into a log or record file. A collector's function is similar to the functions of auditing, logging, and monitoring. A collector, like any auditing system, needs sufficient space on a storage device to record the data it collects. Such data should be treated as more sensitive than the original data, programs, or systems it was collected from. A collector should be placed where it has the ability to review and retrieve information on the system, systems, or network that it is intended to monitor.

CORRELATION ENGINE: A type of analysis system that reviews the content of log files or live events. It is programmed to recognize related events, sequential occurrences, and activity patterns that are interdependent in order to detect suspicious or violating events. A correlation engine does not need to be in line on the network or installed directly onto monitored systems. It must have access to the recorded logs or the live activity stream in order to perform its analysis. This could allow it to operate on or near a data warehouse or centralized logging server or off a switch SPAN port.

FILTER: Used to recognize or match an event, address, activity, content, or keyword and trigger a response. In most cases a filter is used to block or prevent unwanted activities or data exchanges. The most common example of a filtering tool is a firewall.

PLACEMENT ANALYSIS

A filter should be located in line along any communication path where control of data communications is necessary. Keep in mind that filters cannot inspect encrypted traffic, so filtering of such traffic must be done just before encryption or just after decryption.

The placement of a proxy should be between source or origin devices and their destination systems. The location of a transparent proxy must be along the routed path between source and destination, whereas a nontransparent proxy can be located along an alternate or indirect routed path, since source systems will direct traffic to the proxy themselves.

The placement of a firewall should be at any transition between network segments where there is any difference in risk, sensitivity, security, value, function, or purpose. It is a standard security practice to deploy a hardware security firewall between an internal network and the Internet as well as between the Internet, DMZ, extranet, and intranet.

A VPN concentrator should be located on the boundary or border of the organizational network at or near the primary Internet connection. The VPN concentrator may be located inside or outside of the primary network appliance firewall. If the security policy is that all traffic is filtered entering the private network, then the VPN concentrator must be located outside the firewall. If traffic from remote locations over VPNs is trusted, then the VPN concentrator can be located inside the firewall.

An SSL / TLS accelerator should be located at the boundary or border of the organizational network at or near the primary Internet connection and before the resource server being accessed by those protected connections. Usually the SSL/TLS accelerator is located in line with the communication pathway so that no abusive network access can reach the network segment between the accelerator and the resource host. The purpose of this device or service is to offload the computational burden of encryption in order for a resource host to devote its system resources to serving visitors.

A load balancer should be located in front of a group of servers, often known as a cluster, which all support the same resource. The purpose of the load balancer is to distribute the workload of connection requests among the members of the cluster group, so this determines its location.

DDOS MITIGATOR: This will attempt to differentiate legitimate packets from malicious packets. Benign traffic will be sent toward its destination, whereas abusive traffic will be discarded. A DDoS mitigator should be positioned in line along the pathway into the intranet, DMZ, and extranet from the Internet. This provides the DDoS mitigator with the ability to filter all traffic from external attack sources before it reaches servers or the network as a whole.

AGGREGATION SWITCH: The main or master switch used as the interconnection point for numerous other switches. In large network deployments, an initial master primary switch is deployed near the demarcation point, and then additional switches for various floors, departments, or network segments are connected off the master primary switch.

TAP: A tap is a means to eavesdrop on network communications. In the past, taps were physical connections to the copper wires themselves, often using a mechanical means to strip or pierce the insulation to make contact with the conductors. A tap should be installed wherever traffic monitoring on a specific cable is required and when a port mirroring function is either not available or undesired.

> NOTE: Today, taps are often optical or passive and are simply configurations made to copy traffic in multiple directions simultaneously.

SDN / SAN

SOFTWARE-DEFINED NETWORKING: SDN is a unique approach to network operation, design, and management. SDN offers a new network design that is directly programmable from a central location, is flexible, is vendor neutral, and is based on open standards. Using SDN frees an organization from needing to purchase devices from a single vendor.

STORAGE AREA NETWORK: SAN is a secondary network used to consolidate and manage various storage devices. SANs are often used to enhance networked storage devices such as hard drives, drive arrays, optical jukeboxes, and tape libraries so they can be made to appear to servers as if they were local storage. SANs can offer greater storage isolation through the use of a dedicated network. This makes directly accessing stored data difficult and forces all access attempts to operate against a server's restricted applications and interfaces.

SECURE SYSTEM DESIGNS & SECURE STAGING

HARDWARE/FIRMWARE SECURITY

FULL-DISK ENCRYPTION: FDE (also called whole-disk encryption) is often used to provide protection for an OS, its installed applications, and all locally stored data. FDE encrypts all of the data on a storage device with a single master symmetric encryption key. To maximize the defensive strength of full-disk encryption, you should use a long, complex passphrase to unlock the system upon bootup. This passphrase shouldn't be written down or used on any other system or for any other purpose. Whenever the system isn't actively in use, it should be powered down and physically locked against unauthorized access or theft. Hard drive encryption should be viewed as a delaying tactic, rather than as a true prevention of access to data stored on the hard drive.

SELF-ENCRYPTING DRIVE: An SED is now a common term referring to HDDs / SSDs with built-in full-disk encryption. Most of these solutions are proprietary and don't disclose their methods or algorithms, and some have been cracked with relatively easy hacks.

USB ENCRYPTION: Usually related to USB storage devices, which can include both USB-connected hard drives as well as USB thumb drives. If encryption features aren't provided by the manufacturer of a USB device, you can usually add them through a variety of commercial or open-source solutions. One of the best-known, respected, and trusted open-source solutions is VeraCrypt. This tool can be used to encrypt files, folders, partitions, drive sections, or whole drives, whether internal, external, or USB.

TRUSTED PLATFORM MODULE: TPM is both a specification for a cryptoprocessor and the chip in a mainboard supporting this function. A TPM chip is used to store and process cryptographic keys for a hardware-supported/implemented hard drive encryption system. When TPM-based whole-disk encryption is in use, the user/operator must supply a password or physical USB token device to the computer to authenticate and allow the TPM chip to release the hard drive encryption keys into memory. Although this seems similar to a software implementation, the primary difference is that if the hard drive is removed from its original system, it can't be decrypted. Only with the original TPM chip can an encrypted hard drive be decrypted and accessed. With software-only hard drive encryption, the hard drive can be moved to a different computer without any access or use limitations.

HARDWARE SECURITY MODULE: HSM is a special-purpose cryptoprocessor used for a wide range of potential functions. The functions of an HSM can include accelerated cryptography operations, managing and storing encryption keys, offloading digital signature verification, and improving authentication. An HSM can be a chip on a motherboard, an external peripheral, a network-attached device, or an add-on or extension adapter or card. Often an HSM includes tamper protection technology in order to prevent or discourage abuse and misuse even if physical access is obtained by the attacker.

BASIC INPUT/OUTPUT SYSTEM: BIOS is the basic low-end firmware or software embedded in the hardware's electrically erasable programmable read-only memory, EEPROM. The BIOS identifies and initiates the basic system hardware components, such as the hard drive, optical drive, video card, and so on, so that the bootstrapping process of loading an OS can begin. A replacement or improvement to BIOS is Unified Extensible Firmware Interface (UEFI).

UEFI: Provides support for all of the same functions as BIOS with many improvements, such as support for larger hard drives, faster boot times, enhanced security features, and even the ability to use a mouse when making system changes. UEFI also includes a CPU-independent architecture, a flexible pre-OS environment with networking support, secure boot, and forward and backward compatibility. It also runs CPU-independent drivers.

SECURE BOOT: A feature of UEFI that aims to protect the operating environment of the local system by preventing the loading or installing of device drivers or an operating system that is not signed by a preapproved digital certificate. Secure Boot ensures that only drivers and operating systems that pass attestation are allowed to be installed and loaded on the local system. Although the security benefits of Secure Boot attestation are important and beneficial to all systems, there is one important drawback to consider: if a system has a locked UEFI Secure Boot mechanism, it may prevent the system's owner from replacing the operating system or block them from using third-party vendor hardware that has not been approved by the motherboard vendor. If there is any possibility of using alternate OSs or changing hardware components of a system, be sure to use a motherboard from a vendor that will provide unlock codes/keys to the UEFI Secure Boot.

SUPPLY CHAIN SECURITY: The concept that most computers are not built by a single entity. Any finished system has a long and complex history, known as its supply chain, which enabled it or caused it to come into existence. A secure supply chain is one in which all of the vendors or links in the chain are reliable, trustworthy, reputable organizations that disclose their practices and security requirements to their business partners. The goal of a secure supply chain is to ensure that the finished product is of sufficient quality, meets performance and operational goals, and provides stated security mechanisms, and that at no point in the process was any element subjected to unauthorized or malicious manipulation or sabotage.

HARDWARE ROOT OF TRUST: This is based or founded on a secure supply chain. The security of a system is ultimately dependent upon the reliability and security of the components that make up the computer as well as the process it went through to be crafted from original raw materials. If the hardware that is supporting an application has security flaws or a backdoor, or fails to provide proper HSM-based cryptography functions, then the software is unable to accommodate those failings. Only if the root of the system—the hardware itself—is reliable and trustworthy can the system as a whole be considered trustworthy. System security is a chain of many interconnected links; if any link is weak, then the whole chain is untrustworthy.

ELECTROMAGNETIC INTERFERENCE: EMI is the noise caused by electricity when used by a machine or flowing along a conductor. EMI shielding is important for network-communication cables as well as for power- distribution cables. EMI shielding can include upgrading from UTP, unshielded twisted pair, to STP, shielded twisted pair, running cables in shielding conduits, or using fiber-optic networking cables. Generally, these two types of cables, networking and electrical, should be run in separate conduits and be isolated and shielded from each other. The strong magnetic fields produced by power-distribution cables can interfere with network-communication cables.

OPERATING SYSTEMS

This section includes several specific examples of OS types, labels, and groupings. In all cases, the selection of an OS should focus on features and capabilities without overlooking the native security benefits. Although security can often be added through software installation, native security features are often superior.

NETWORK OPERATING SYSTEM: An NOS is any OS that has native networking capabilities and was designed with networking as a means of communication and data transfer. Most OSs today are NOSs, but not all OSs are network capable. There are still many situations where a stand-alone or isolated OS is preferred for function, stability, and security.

SERVER: A form of NOS that is a resource host offering data, information, or communication functions to other requesting systems. Servers are the computer systems on a network that support and maintain the network. They require greater physical and logical security protections than workstations because they represent a concentration of assets, value, and capabilities. End users should be restricted from physically accessing servers, and they should have no reason to log on directly to a server—they should interact with servers over a network through their workstations.

WORKSTATION: Another form of NOS and is a resource consumer. A workstation is typically where an end user will log in and then from the workstation reach out across the network to servers to access resources and retrieve data. Access to workstations should be restricted to authorized personnel. One method to accomplish this is to use strong authentication, such as two-factor authentication with a smartcard and a password or PIN.

APPLIANCE OS: Yet another variation of NOS. An appliance NOS is a stripped-down or single-purpose OS that is typically found on network devices, such as firewalls, routers, switches, wireless access points, and VPN gateways. An appliance NOS is designed around a primary set of functions or tasks and usually does not support any other capabilities.

KIOSK OS: Can be either a stand-alone OS or a variation of NOS. A kiosk OS is designed for end-user use and access. The end user may be an employee of an organization or anyone from the general public. A kiosk OS is locked down so that only pre-authorized software products and functions are enabled and will revert to the locked-down mode each time it is rebooted. Some will even revert if they experience a flaw, crash, error, or any attempt to perform an unauthorized command or launch an unapproved executable. The goal and purpose of a kiosk OS is to provide a robust information service to a user while preventing accidental or intentional system misuse.

MOBILE OS: An NOS that is designed to operate on a portable device. Although a portable device can be defined as any device with a battery, a mobile OS is designed for portable devices for which traditional NOSs are too large or too resource demanding. A mobile OS is designed to optimize performance of limited resources. It might be designed around a few specific mobile device features, such as phone calls, text messages, and taking photographs, or it may be designed to support a wide range of user-installed software or applications.

PATCH MANAGEMENT: The formal process of ensuring that updates and patches are properly tested and applied to production systems. Using vendor updates to OSs, applications, services, protocols, device drivers, and any other software is the best way to protect your environment from known attacks and vulnerabilities. Not all vendor updates are security related, but any error, bug, or flaw that can be exploited to result in damaged data, disclosure of information, or obstructed access to resources should be addressed. The output of a patch management tool will be a report indicating that monitored systems are or are not in compliance with the approved updates. Some patch management systems may be able to automatically apply approved patches, whereas others require the administrator to install updates manually.

A key element in securing a system is to reduce its attack surface by disabling unnecessary ports and services. The attack surface is the area that is exposed to untrusted networks or entities and vulnerable to attack. If a system is hosting numerous services and protocols, its attack surface is larger than that of a system running only essential services and protocols. Any unused application service ports should be specifically blocked / disabled. Port or interface disabling is a physical option that renders a connection port electrically useless. Port blocking is a service provided by a software or hardware firewall that blocks or drops packets directed toward disallowed ports. You must also determine which services are essential on your specific system. Services that are essential on a web server may not be essential on a file server or an email server.

LEAST FUNCTIONALITY: This is when you always select and install the solution with the least functionality, without any unnecessary additional capabilities and features. You will likely have a more secure result than opting for any solution with more options than necessary. This is another perspective on minimizing your attack surface. Rather than removing and blocking components that are unneeded or unwanted, select hardware and software systems that have minimal additional capabilities beyond what is strictly needed for the business function or task. All company systems should be operating within expected parameters and compliant with a defined baseline of Secure Configuration. Any system that is determined to be out of baseline should be removed from the production network in order to investigate the cause. If the deviation was due to normal work-related actions and activities, it may be necessary to update the baseline and/or implement more restrictive system modification policies, such as whitelisting or using static systems. A static system is an environment in which users cannot make changes or the few changes users can implement are only temporary and are discarded once the user logs out.

TRUSTED OS: An access-control feature requiring a specific OS to be present in order to gain access to a resource. By limiting access to only those systems that are known to implement specific security features, resource owners can ensure that security violations will be less likely. Examples of trusted OSs include Trusted Solaris, Apple macOS X, HP-UX, and AIX.

APPLICATION WHITELISTING: A security option that allows authorized software to execute and blocks everything else by default. Application Blacklisting allows software execution by default and denies it by exception. Application whitelisting may produce logs regarding the attempts of users to execute software that is not included on the approved list. Analyzing and interpreting this output can help determine whether additional software needs to be approved or whether users are attempting to perform unauthorized tasks for personal benefit or attempt work tasks for which they are not trained, skilled, or authorized.

If you don't need it, don't keep it. This may be an optional mantra for you in real life, but in terms of security, it's the first of two — the second is, *lock down what's left.* Getting rid of unnecessary services and accounts is just the beginning of proper security and environment hardening. Leaving behind default / unused accounts gives hackers more potential points to compromise.

PERIPHERALS

WIRELESS KEYBOARD: These connect to computers over Bluetooth, WiFi, or some other radio wave-based communication. In most cases, wireless keyboards do not use encryption, so any listening device within range may be able to eavesdrop on the characters typed into a wireless keyboard. Do not use wireless keyboards for sensitive systems or when you are typing sensitive, confidential, or valuable information.

WIRELESS MOUSE: Like wireless keyboards, these connect to a computer via Bluetooth, WiFi, or some other radio wave-based communication. In most cases, a wireless mouse do not use encryption, so any listening device within range may be able to eavesdrop on the movements and activities of a wireless mouse. Do not use a wireless mouse on sensitive systems.

SYSTEM DISPLAY: Your display can show sensitive, confidential, or personal information. It is important to orient your system display so that it is hard to see unless you are sitting or in your work position. You do not want others in the general area or who just walk by to be able to see the contents of your screen. It may be worthwhile to install screen filters, also called privacy filters.

PORTABLE DEVICES: Many of these that do not have native wireless support can be enhanced using a microSD or SD card with WiFi. Such memory storage cards include their own WiFi adapter, which can in turn provide wireless connectivity to the mobile device. When these cards are used in a device that already has networking capabilities, the device can serve as an additional attack path for hackers, especially if it automatically connects to plain-text WiFi networks. In general, avoid the use of any WiFi-enabled storage expansion card.

PRINTER: Many printers are network attached, meaning they can be directly connected to the network without being directly attached to a computer. A network-attached printer serves as its own print server. It may connect to the network via cable or through wireless. Some devices are more than just printers and may include fax, scanning, and other functions. These are known as multifunction devices, MFDs. Any of these MFDs connected to a network can be a potential breach point. This may be due to flaws in the firmware of the device as well as whether or not the device uses communication encryption.

UNIVERSAL SERIAL BUS: USB devices are ubiquitous today. Nearly every worker who uses a computer possesses a USB storage device, and most portable devices connect via USB. To protect against USB threats, the only real option is to fully disallow the use of USB devices and lock down all USB ports. Some organizations not only disable USB functionality, but also physically fill USB ports with silicon, epoxy, or a similar material to ensure that USB devices can't be used.

DIGITAL CAMERAS: These can become a security risk when they are used to take photographs of sensitive documents or computer screens. A digital camera can serve as a storage device when connected via a cable to a computer system, thus allowing the user to transmit confidential files outside the organization or to bring in malware from outside. Most digital cameras include GPS chips in order to geotag photos and videos created on the device. This can reveal sensitive or secret locations, as well as the time and date of a photo being taken or a video being recorded.

SANDBOXING

This is a means of quarantine or isolation. It is simple to implement in a virtualization context because you can isolate a virtual machine with just a few mouse clicks or commands. Once the suspect code is deemed safe, you can release it to integrate with the environment. If it's found to be malicious, unstable, or otherwise unwanted, it can quickly be removed from the environment with little difficulty.

ENVIRONMENT

DEVELOPMENT NETWORK: Where new software code is being crafted by on-staff programmers and developers. For some organizations, this might also be where custom-built hardware is being created. This network is to be fully isolated from all other network divisions in order to prevent ingress of malware or egress of unfinished products.

TEST NETWORK: Where in-development products or potentially final versions of products are subjected to a battery of evaluations, stress tests, vulnerability scans, and even attack attempts in order to determine whether the product is stable, secure, and ready for deploying into the production network.

STAGING NETWORK: Where new equipment, whether developed in-house or obtained from external vendors, is configured to be in compliance with the company's security policy and configuration baseline. Once a system has been staged, it can be moved to the test network for evaluation. After passing evaluation, it can be deployed into the production network.

PRODUCTION NETWORK: Where the everyday business tasks and work processes are accomplished. It should only be operating on equipment and systems that have been properly staged and tested. The production network should be managed so that it is not exposed to the risk and unreliability of new systems and untested solutions. The goal of the production network is to support the confidentiality, integrity, and availability of the organization's data and business tasks.

SECURE BASELINE

SECURITY POSTURE: This is the level to which an organization is capable of withstanding an attack. An organization may have good or poor posture. A plan and implementation are parts of the security posture often known as the secure baseline or security baseline. These include detailed policies and procedures, implementation in the IT infrastructure and the facility, and proper training of all personnel.

The basic procedure for establishing a security baseline or hardening a system is as follows:

- Remove unneeded components, such as protocols, applications, services, and hardware, including device drivers.
- Update and patch the OS and all installed applications, services, and protocols.
- Configure all installed software as securely as possible.
- Impose restrictions on information distribution for the system, its active services, and its hosted resources.

SECURITY TEMPLATE: A set of security settings that can be mechanically applied to a computer to establish a specific configuration. Security templates can be used to establish baselines or bring a system up to compliance with a security policy. Once a security template exists, you can use it to configure a new or existing machine, or to compare the current configuration to the desired configuration.

OPERATING SYSTEM HARDENING: The process of reducing vulnerabilities, managing risk, and improving the security provided by or for an OS. This is usually accomplished by taking advantage of an OS's native security features and supplementing them with add-on applications such as firewalls, antivirus software, and malicious-code scanners.

FILESYSTEM: The type of filesystem in use greatly affects the security offered by that system. A filesystem that incorporates security, such as access control and auditing, is a more secure choice than one without incorporated security. One great example of a secured file system is the Microsoft New Technology File System, NTFS. It offers file and folder level access permissions and auditing capabilities.

INTEGRITY MEASUREMENT

The primary means of integrity measurement or assessment is the use of a hash. Hashing is a type of cryptography that isn't an encryption algorithm. Instead, hashing is used to produce a unique identifier—known as a hash value, hash, checksum, message authentication code, fingerprint, thumbprint, or message digest—of data. A hash serves as an ID code to detect when the original data source has been altered, since the altered file will produce a different hash value. The data could be a file, a hard drive, a network traffic packet, or an email message. The hash value is used to detect when changes have been made to a resource. In other words, hashing is used to detect violations of data integrity.

APPLICATION SECURITY

SOFTWARE DEVELOPMENT LIFE CYCLE MODELS

Two of the dominant SDLC concepts are the waterfall model and the Agile model.

WATERFALL: This model consists of seven stages, or steps. A recent revision of the waterfall model allows for some movement back into earlier phases for addressing oversights or mistakes discovered in later phases. The primary criticism of the waterfall model is the limitation to only return to the immediately previous phase. It forces the completion of a product that is known to be flawed or not to fulfill goals.

AGILE: An encompassing term for a number of iterative and incremental approaches to creating products — iterative because the team revisits the product, and incremental because the team completes features as it works. Agile processes promote sustainable development. Sponsors, developers, and users should be able to maintain a constant pace indefinitely.

DevOps is a combination of practices, procedures, and culture that marry software development and IT operations to accelerate the delivery of high quality software releases while maintaining stability and functionality. Security automation is important to DevOps in order to ensure that issues and vulnerabilities are discovered early and properly addressed before product release. This will include automating vulnerability scans and code attacks against preproduction code, using fuzzing techniques to discover logic flaws or the lack of input sanitization, and using code scanners that evaluate software for flaws and input management mistakes.

In order for security to be successful in any development endeavor, it must be integrated at the beginning and maintained throughout the development process. Secure DevOps must adopt a continuous integration approach to ensure that automated tools, automated testing, and manual injection of security elements are included throughout the process of product development.

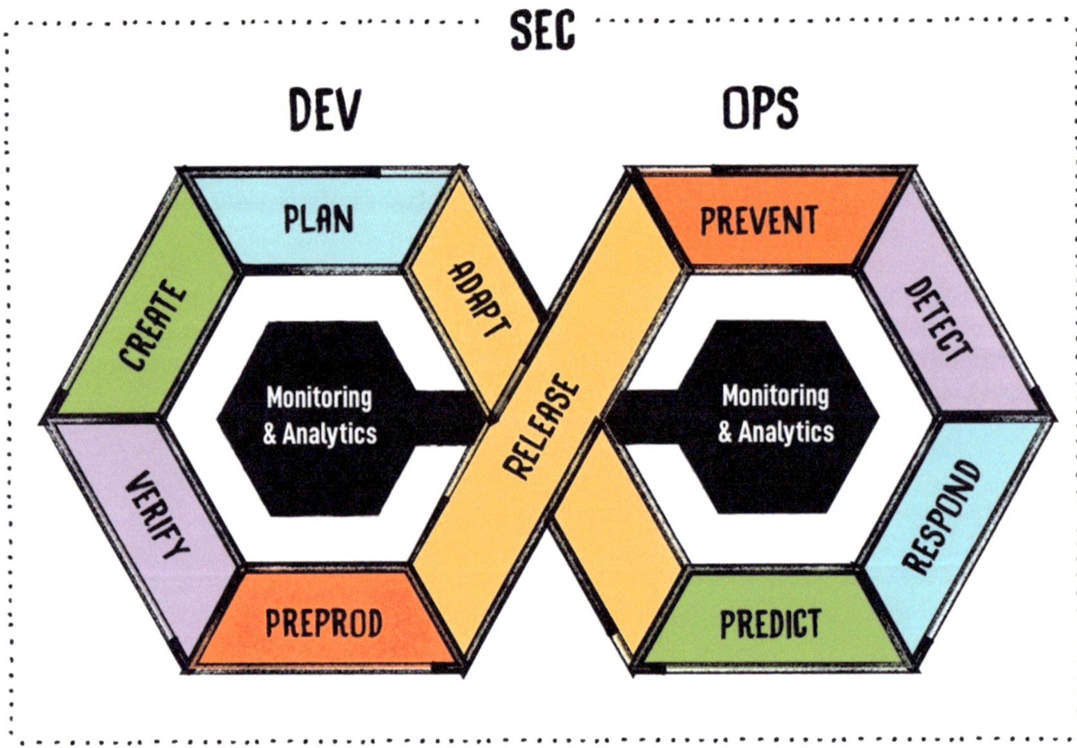

BASELINE: A minimum level of function, response, and security that must be met in order for the project to proceed toward release. Any software product that does not meet or exceed the baseline is rejected and must return to development to be improved. A baseline is a form of quality control. Without quality control, neither the needs of the customer nor the reputation of the vendor are respected.

IMMUTABLE SYSTEM: A server or software product that, once configured and deployed, is never altered in place. Instead, when a new version is needed or a change is necessary, a revised version is crafted and the new system is then deployed to replace the old one. Immutable systems ensure that each member of the server group is exactly the same, so when something needs to change, it is first developed and tested in a staging area, and then, when finalized, the new version fully replaces the previous version.

INFRASTRUCTURE AS CODE: A change in how hardware management is perceived and handled. It is viewed as just another collection of elements to be managed in the same way that software and code are managed under DevOps. This alteration in hardware management approach has allowed many organizations to streamline infrastructure changes so that they occur more easily, more rapidly, more securely and safely, and more reliably than before.

VERSION CONTROL & CHANGE MANAGEMENT

Version control and change management are critical duties for maintaining secure and functioning code. Version control creates point in time references to files that can be used to quickly identify bugs, organize major feature changes, or critical patching. Version control commands a form of accountability for code authors and security staff by recording change authors and change justification. Finally, version control provides a form of safety net for critical codebases and systems that permit for operational backups in case of emergency.

Change management goes hand in hand with version control in maturing an organization's security posture. Change management dictates a set of processes and procedures that can be used to effectively implement change in an organization, while ensuring key stakeholders are informed and prepared for all changes. By effectively implementing change management, stakeholders are able to prepare as best as possible for any impact the change may have. Proper change management not only helps offset the consequences of Murphy's Law, but also aids in building cross department relationships by making security a collaborative effort, instead of a road block.

PROVISIONING & DEPROVISIONING

The processes of provisioning and deprovisioning are most commonly associated with identity management and the act of account creation and permission management. The provisioning process can often be long and tedious for many organizations if key stakeholders do not agree on a process that provides automation and documentation for all. By working closely with development and operations teams, security groups can ensure users are created in a timely matter, with standard security controls that permit quick onboarding and limited deviation from policy. Standardization of provisioning also allows more flexibility for unique use cases, such as cloud developer accounts, and for insight into any abnormal account creation.

While the consequences for limited security in provisioning can mean a delay in account creation, the consequences for limited security in deprovisioning can be much more severe. The act of deprovisioning is frequently associated with permission removal and account deletion. Deprovisioning can occur when an individual changes job roles and no longer needs administrative access to a database, or when a user leaves the company and their active directory account must be turned off entirely. Failure to securely deprovision can leave large holes for attackers to exploit with a myriad of attacks, such as password stuffing, social engineering, or extortion. Similar to provisioning, security should work with all groups involved in account deletion to automate and document deprovisioning activities. For sensitive or VIP users, security may be more involved in the deprovisioning process with potential heightened monitoring or forensic imaging.

SECURE CODING TECHNIQUES

FAIL-SECURE DESIGN: When errors occur, the program falls back to a secure state. However, the programmer must code this into the application in order for a true fail-secure response to take place. This should include error and exception handling. When a process, a procedure, or an input results in or causes an error, the system should revert to a more secure state.

INPUT VALIDATION: An aspect of defensive programming intended to ward off a wide range of input-focused attacks, such as buffer overflows, command or code injection. Input validation checks all input received before it's allowed to be processed.

NORMALIZATION: A database programming and management technique used to reduce redundancy. The goal of normalization is to prevent redundant data, which is a waste of space and can also increase processing load. A normalized database is more efficient and can allow for faster data retrieval operations.

STORED PROCEDURE: This is a subroutine or software module that can be called on or accessed by applications interacting with a relational database management system, RDBMS. The stored procedures may be used for data validation during input, managing access control, assessing the logic of data, and more. Stored procedures can make some database applications more efficient, consistent, and secure.

CODE SIGNING: The activity of crafting a digital signature of a software program in order to confirm that it was not changed and who it is from.

ENCRYPTION: Should be used to protect data in storage and data in transit. Programmers should adopt trusted and reliable encryption systems into their applications.

OBFUSCATION: Also called camouflage, this is the practice of crafting code specifically so that other programmer will find it difficult to decipher. This technique might be adopted in order to prevent unauthorized parties from understanding proprietary solutions. These techniques can also be adopted by malicious programmers to hide the true intentions and purposes of software.

CODE REUSE: The inclusion of preexisting code in a new program, used as a way to quicken the development process by adopting and reusing existing code.

DEAD CODE: Any section of software that is executed but whose output or result is not used by any other process. Effectively the execution of dead code is a waste of time and resources.

SERVER-SIDE VALIDATION: This is suited for protecting a system against input submitted by a malicious user. Client-side validation is also important, but it is focused on providing better responses or feedback to the typical user. Although all validation can take place on the server side, it is often a more complex process and introduces delays to the interaction. A combination of server-side and client-side validation allows more efficient interaction while maintaining reasonable security defenses.

MEMORY MANAGEMENT: Programmers should include code in their software that focuses on proper memory management. Software should pre-allocate memory buffers but also limit the input sent to those buffers. Some programmers focus on getting new code to function with the intention of returning to the code in the future to improve security and efficiency. Unfortunately, if the functional coding efforts take longer than expected, it can result in the security revisions being minimized or skipped. Always be sure to use secure coding practices, such as proper memory management, to prevent a range of common software exploitations, such as buffer overflow attacks.

SOFTWARE DEVELOPMENT KITS: SDKs are often essential tools for a programmer, along with third-party software libraries. This can allow programmers to focus on their custom code and logic. However, when you are using third-party software libraries, the pre-crafted code may include flaws, backdoors, or other exploitable issues that are unknown and yet undiscovered.

UNAUTHORIZED DATA EXPOSURE: When software does not adequately protect the data it processes. Programmers need to include authorization, authentication, and encryption schemes in their products in order to protect against data leakage, loss, and exposure.

CODE QUALITY & TESTING

STATIC CODE ANALYZERS: Review the raw source code of a product without executing the program. This debugging effort is designed to locate flaws in the software code before the program is run on a target or customer system. Static code analysis is often a first step in software quality and security testing.

DYNAMIC ANALYSIS: The testing and evaluation of software code while the program is executing. The executing code is then subjected to a range of inputs to evaluate its behavior and responses. One method of performing dynamic analysis is known as fuzzing. Fuzzing is a software-testing technique that generates inputs for targeted programs. The goal of fuzzing is to discover input sets that cause errors, failures, and crashes, or to discover other unknown defects in the targeted program. Once a fuzzing tool discovers a constructed input that causes an abnormal behavior in the target application, the input and response are recorded into a log. The log of interesting inputs is reviewed by a security professional or a hacker. With the right skills and tools, the results of fuzzing can be transformed into a patch that fixes discovered defects or an exploit that takes advantage of them.

STRESS TESTING: Another variation of dynamic analysis in which a hardware or software product is subjected to various levels of workload to evaluate its ability to operate and function under stress. Stress testing can start with a modest level of traffic and increase to abnormally high levels. The purpose of stress testing is to gain an understanding of how a product will perform, react, or fail in various circumstances between normal conditions and DoS level traffic or load.

SANDBOX: A software implementation of a constrained execution space used to contain an application. Sandboxing is often used to protect the overall computer from a new, unknown, untested application. The sandbox provides the contained application with direct or indirect access to sufficient system resources to execute, but not the ability to make changes to the surrounding environment or storage devices, beyond its own files. Sandboxing is commonly used for software testing and evaluating potential malware, and it is the basis for the concept of virtualization

MODEL VERIFICATION: Part of the software development process often used to ensure that the crafted code remains in compliance with a development process, an architectural model, or design limitations. Model verification can also extend to ensuring that a software solution is able to achieve the desired real-world results by performing operational testing. Model verification can ensure that a product maintains compliance with security baseline requirements during development

COMPILED VS RUNTIME CODE

If the code is converted to machine language using a compiler crafting an output executable, then the language is described as compiled. The resulting executable file can be run at any time. If the code remains in its original human-readable form and is converted into machine language only at the moment of execution, the language is a runtime compiled language. Compiled code is harder for an attacker to inject malware into, but it is harder to detect such malware. Runtime code is easier for an attacker to inject malware into, but it is easier to detect such malware.

ARCHITECTURE, DESIGN, & IDENTITY MANAGEMENT

CLOUD, VIRTUALIZATION, & AUTOMATION

HYPERVISOR

The Hypervisor, also known as the virtual machine monitor (VMM), is the component of virtualization that creates, manages, and operates the virtual machines. The computer running the hypervisor is known as the host OS, and the OSs running within a hypervisor-supported virtual machine are known as guest OSs.

TYPE I HYPERVISOR: A native or bare-metal hypervisor. In this configuration, there is no host OS; instead, the hypervisor installs directly onto the hardware in place of the host OS. Type 1 hypervisors are often used to support server virtualization. This allows for maximization of the hardware resources while eliminating any risks or resource reduction caused by a host OS.

TYPE II HYPERVISOR: A hosted hypervisor. In this configuration, a standard OS, like Microsoft Windows, is present on the hardware, and then the hypervisor is installed as another software application. Type II hypervisors are often used in relation to desktop deployments, where the guest OSs offer safe sandbox areas to test new code, allow the execution of legacy applications, support apps from alternate OSs, and provide the user with access to the capabilities of a host OS.

Another variation of virtualization is focusing on applications instead of entire operating systems. Application cells or Application Containers are used to virtualize software so they can be ported to almost any OS.

VM SPRAWL AVOIDANCE

VM sprawl occurs when an organization deploys numerous virtual machines without an overarching IT management or security plan in place. Uncontrolled VM creation can quickly lead to a situation where manual oversight cannot keep up with system demand. To prevent or avoid VM sprawl, a policy for developing and deploying VMs must be established and enforced. This should include establishing a library of initial or foundation VM images that are to be used to develop and deploy new services.

VM ESCAPE PROTECTION

VM escaping occurs when software within a guest OS is able to breach the isolation protection provided by the hypervisor in order to violate the container of other guest OSs or to infiltrate a host OS. VM escaping can be a serious problem, but steps can be implemented to minimize the risk. First, keep highly sensitive systems and data on separate physical machines. Second, keep all hypervisor software current with vendor-released patches. Third, monitor attack, exposure, and abuse indexes for new threats to your environment.

CLOUD STORAGE

Cloud storage is the idea of using storage capacity provided by a cloud vendor as a means to host data files for an organization. Cloud storage can be used as a form of backup or support for online data services. Cloud storage may be cost effective, but it is not always high speed or low latency.

CLOUD DEPLOYMENT MODELS

Cloud computing is a popular term that refers to performing processing and storage elsewhere, over a network connection, rather than locally. Cloud computing is often thought of as Internet-based computing.

SOFTWARE AS A SERVICE: SaaS is a derivative of Platform as a Service. It provides on-demand online access to specific software applications or suites without the need for local installation.

PLATFORM AS A SERVICE: PaaS is the concept of providing a computing platform and software solution stack to a virtual or cloud-based service. Essentially, it involves paying for a service that provides all the aspects of a platform. A PaaS solution grants the customer the ability to run custom code of their choosing without needing to manage the environment.

INFRASTRUCTURE AS A SERVICE: IaaS takes the platform as a service model another step forward and provides not just on-demand operating solutions but complete outsourcing options. Infrastructure as a Service allows an enterprise to quickly scale up new software- or data-based services/solutions through cloud systems without having to install massive hardware locally.

PRIVATE CLOUD: A cloud service that is within a corporate network and isolated from the Internet. The private cloud is for internal use only. An organization can outsource its private cloud to an external provider.

PUBLIC CLOUD: A cloud service that's accessible to the general public, typically over an internet connection. An organization's or individual's data is usually separated from other customers' data in a public cloud, so the overall purpose / usage of the cloud is the same for all customers.

HYBRID CLOUD: A mixture of private and public cloud components. For example, an organization could host a private cloud for exclusive internal use but distribute some resources onto a public cloud for the public, business partners, customers, the external sales force, and so on.

COMMUNITY CLOUD: Maintained, used, and paid for by a group of users or organizations for their shared benefit, such as collaboration and data exchange. This may allow for some cost savings compared to accessing private or public clouds independently.

ON-PREMISE VS HOSTED VS CLOUD

ON-PREMISE: This is the traditional deployment solution in which an organization owns the hardware, licenses the software, and operates and maintains the systems on its own, usually within their own building.

CLOUD: In this deployment concept, an organization contracts with a third-party cloud provider. The cloud provider owns, operates, and maintains the hardware and software. The organization pays a monthly fee (often based on a per-user multiplier) to use the cloud solution.

HOSTED: The organization must license software, then operate and maintain it. The hosting provider owns, operates, and maintains the hardware that supports the organization's software.

On-Premise solutions do not have ongoing monthly costs, but may be more costly because of initial up-front costs of obtaining hardware and licensing. On-Premise solutions offer full customization, provide local control over security, do not require Internet connectivity, and provide local control over updates and changes. However, they also require significant administrative involvement for updates and changes, require local backup and management, and are more challenging to scale.

Cloud solutions often have lower up-front costs, lower maintenance costs, vendor-maintained security, and scalable resources, and they usually have high levels of uptime and availability from anywhere. However, cloud solutions do not offer customer control over OS and software, such as updates and configuration changes; offer minimal customization; and are often inaccessible without Internet connectivity. In addition, the security policies of the cloud provider might not match those of the organization.

VDI / VDE

Virtual Desktop Infrastructure, VDI, is a means to reduce the security risk and performance requirements of end devices by hosting virtual machines on central servers that are remotely accessed by users. VDI has been adopted for mobile devices and has already been widely used on tablets and notebook computers.

CLOUD ACCESS SECURITY BROKER

A cloud access security broker, CASB, is a security policy enforcement solution that may be installed on premise or in the cloud. The goal of a CASB is to enforce proper security measures and ensure that they are implemented between a cloud solution and a customer organization.

SECURITY AS A SERVICE

Security as a Service, SECaaS, is a cloud provider concept in which security is provided to an organization through or by an online entity. The purpose of an SECaaS solution is to reduce the cost and overhead of implementing and managing security locally.

AUTOMATION / SCRIPTING

Automation is the control of systems on a regularly scheduled, periodic, or triggered basis that does not require manual hands-on interaction. Automation includes concepts such as scheduled backups, archiving of log files, blocking of failed access attempts, and blocking communications based on invalid content in initial packets or due to traffic seeming like a port scan.

Automated Courses of Action ensure that a specific series of steps or activities are performed in the correct order each and every time. This helps ensure consistency of results, which in turn establishes consistent security.

In order for security monitoring to be effective, it must be continuous in several ways. First, it must always be running and active. Second, security monitoring should be continuous across all user accounts, not just end users. Third, security monitoring should be continuous across the entire IT infrastructure. Fourth, security monitoring should be continuous for each user from the moment of attempted logon until the completion of a successful logoff or disconnect.

Automation is only effective if accurate. Repeating execution of a flawed program may leave the environment with reduced security rather than improved or maintained security. All systems need to have a defined baseline of configuration that is clearly documented.

TEMPLATES

A Template is a pre-established starting point. A template can be crafted for a plethora of concerns in an environment, including a security policy, a procedure, a contract, a submission form, a system image, a software configuration, and a firewall rule set.

MASTER IMAGE

A Master Image or Gold Master is a crafted setup and configuration of a software product or an entire computer system. A Master Image is created just after the target system has been manually installed, patched, and configured. A Master Image is employed to quickly roll out new versions of a system.

NON-PERSISTENCE

A Non-Persistent system is a computer system that does not allow, support, or retain changes. Thus, between uses and/or reboots, the operating environment and installed software are exactly the same.

SNAPSHOT: A copy of the live current operating environment. Snapshots are mostly known relative to virtual machines and guest OSs. However, the term can be loosely employed to refer to any system-wide backup that can be restored to a previous state or condition of configuration and operation.

REVERT TO KNOWN STATE: A type of backup or recovery process. Many databases support a known state reversion for returning to a state of data before edits or changes were implemented.

ROLLBACK TO KNOWN CONFIGURATION: A concept similar to that of reverting to a known state, however state reversion may address a larger portion of the environment than just configuration. A known configuration is just a collection of settings, not likely to include any software elements, such as code present before a patch was applied.

LIVE BOOT MEDIA: A portable storage device that can be used to boot a computer. Live boot media contains a read-to-run or portable version of an operating system.

ELASTICITY

Elasticity is the ability of a system to adapt to workload changes by allocating or provisioning resources in an automatic responsive manner. Elasticity is a common feature of cloud computing, where additional system resources or even additional hardware resources can be provisioned to a server when demand for its services increases.

SCALABILITY

Scalability is the ability for a system to handle an ever-increasing level or load of work. It can also be the potential for a system to be expanded to accommodate future growth. Some amount of additional capacity can be implemented into a system so it can take advantage of the dormant resources automatically as need demands.

DISTRIBUTIVE ALLOCATION

Distributive or Distributed Allocation is the concept of provisioning resources across multiple servers or services as needed, rather than pre-allocating or concentrating resources based exclusively on physical system location. This is a form of load balancing but with a focus on the supporting resources rather than the traffic or request load.

REDUNDANCY

Redundancy is the implementation of secondary or alternate solutions. Commonly, redundancy refers to having alternate means to perform work tasks or accomplish IT functions. Redundancy helps reduce single points of failure and improves fault tolerance. When there are multiple pathways, copies, devices, and so on, there is reduced likelihood of downtime if something fails.

REDUNDANT SERVER: A mirror or duplicate of a primary server that receives all data changes immediately after they are made on the primary server. In the event of a failure of the primary server, the secondary or redundant server can immediately take over and replace the primary server in providing services to the network.

FAULT TOLERANCE

Fault tolerance is the ability of a system to handle or respond to failure smoothly. This can include software, hardware, or power failure.

HIGH AVAILABILITY

High Availability means the availability of a system has been secured to offer very reliable assurance that the system will be online, active, and able to respond to requests in a timely manner, and that there will be sufficient bandwidth to accomplish requested tasks in the time required. Both of these concerns are central to maintaining continuity of operations. High availability is a form of fault tolerance—or, rather, a benefit of providing reliable fault tolerance. Fault tolerance is the ability of a network, system, or computer to withstand a certain level of failures, faults, or problems and continue to provide reliable service.

RAID

An example of a high-availability solution is a Redundant Array of Independent Disks, or a RAID. A RAID solution employs multiple hard drives in a single storage volume.

- RAID 0 provides performance improvement but not fault tolerance, known as striping; it uses multiple drives as a single volume.
- RAID 1 provides mirroring; meaning the data written to one drive is exactly duplicated to a second drive in real time.
- RAID 5 provides striping with parity: three or more drives are used in unison, and one drive's worth of space is consumed with parity information. The parity information is stored across all drives. If any single drive of a RAID 5 volume fails, the parity information is used to rebuild the contents of the lost drive on the fly. A new drive can replace the failed drive, and the RAID 5 system rebuilds the contents of the lost drive onto the replacement drive. RAID 5 can only support the failure of one disk drive.

PHYSICAL SECURITY CONTROLS

LIGHTING

Lighting is a commonly used form of perimeter security control. The primary purpose of lighting is to discourage casual intruders, trespassers, prowlers, or would-be thieves who would rather perform their misdeeds in the dark, such as vandalism, theft, and loitering.

SIGNS

Signs can be used to declare areas off limits to those who are not authorized, indicate that security cameras are in use, and disclose safety warnings. Signs are useful in deterring minor criminal activity, establishing a basis for recording events, and guiding people into compliance or adherence with rules or safety precautions.

FENCES

A fence is a perimeter-defining device. A gate is a controlled exit and entry point in a fence. A cage is an enclosed fence area that can be used to protect assets from being accessed by unauthorized individuals. Cages can be used inside or outside.

SECURITY GUARDS

Security guards can be posted around a perimeter or inside to monitor access points or watch detection and surveillance monitors. The real benefit of guards is that they're able to adapt and react to various conditions and situations. Guards can learn to recognize attack and intrusion activities and patterns, adjust to a changing environment, and make decisions accordingly.

ALARMS

Alarms or Physical IDSs are systems, whether automated or manual, that are designed to detect an attempted intrusion, breach, or attack; the use of an unauthorized entry point; or the occurrence of some specific event at an unauthorized or abnormal time. IDSs used to monitor physical activity may include security guards, automated access controls, and motion detectors as well as other specialty monitoring techniques.

PHYSICAL SECURITY

Any device or removable media containing highly sensitive information should be kept locked securely in a safe when not in active use. You can install a department-wide safe that is managed by a single person, or you can install per-desk safes. A per-desk safe is often smaller, but it lets workers store devices and documentation securely while also allowing quick access. Long-term storage of media and devices may require safes as well. Safes may be present onsite, or you can contract with an offsite storage facility to provide a safe for secure storage.

Cabinets, device enclosures, rack-mounting systems, patch panels, wiring closets, and other equipment and cable containers can provide additional physical security through the use of locking mechanisms. Locking cabinets and other forms of containers can block or reduce access to power switches, adapter ports, media bays, and cable runs.

PROTECTED DISTRIBUTION

Protective distribution systems, PDSs, also known as protected cabling systems, are the means by which cables are protected against unauthorized access or harm. The goals of PDSs are to deter violations, detect access attempts, and otherwise prevent compromise of cables.

AIRGAP

Airgap is a security control implemented at the network level which aims to secure one or more endpoints from unsecure networks. Frequently this is utilized to protect critical or sensitive systems from the Internet. The control sometimes involves physically securing systems by use of secure rooms or device separation, though the primary focus of airgaps is to remove network interfaces from other networks.

MANTRAP

A Mantrap is a high-security barrier entrance device. It's a small room with two doors; one in the trusted environment and one opening to the outside. The mantrap works like this:
- A person enters the mantrap.
- Both doors are locked.
- The person must properly authenticate to unlock the inner door to gain entry.
- If authentication fails, security is notified while intruder is detained in the mantrap.

FARADAY CAGE

A Faraday cage is an enclosure that blocks or absorbs electromagnetic fields or signals. Faraday cage containers, computer cases, rack-mount systems, rooms, or even building materials are used to create a blockage against the transmission of data, information, metadata, or other emanations from computers and other electronics. Devices inside a Faraday cage can use electromagnetic (EM) fields for communications, such as wireless or Bluetooth, but devices outside the cage will not be able to eavesdrop on the signals of the systems within the cage.

LOCKS

Locks are used to keep doors and containers secured in order to protect assets. Hardware conventional locks and even electronic or smart locks are used to keep specific doors or other access portals closed and prevent entry or access to all but authorized individuals. Doors used to control entrance into secured areas can be protected by locks that are keyed to biometrics. A biometric lock requires that the person present a biometric factor.

BIOMETRICS

Biometrics is the term used to describe the collection of physical attributes of the human body that can be used as identification or authentication factors. Biometrics fall into the authentication factor category of something you are.

BARRICADES

Barricades, in addition to fencing, are used to control both foot traffic and vehicles.

TOKEN DEVICE

A token device or an access card can be used as an element in authentication when gaining physical entry into a facility.

ENVIRONMENTAL MONITORING

Environmental monitoring is the process of measuring and evaluating the quality of the environment within a given structure. This can focus on general or basic concerns, such as temperature, humidity, dust, smoke, and other debris. However, more advanced systems can include chemical, biological, radiological, and microbiological detectors.

HVAC

Heating, ventilating, and air-conditioning, HVAC, management is important for two reasons: temperature and humidity. In the mission-critical server vault or room, the temperature should be maintained around a chosen set point to support optimal system operation.

Hot and cold aisles are a means of maintaining optimum operating temperature in large server rooms. The overall technique is to arrange server racks in lines separated by aisles.

EARLY FIRE DETECTION

Fire is a common problem that must be addressed in the design of any facility. Early fire detection and suppression is important because the earlier the discovery, the less damage is caused to the facility and equipment. Human life and safety are without question the top concerns, but sufficient focus needs to be placed on providing physical security for buildings and other real-world assets.

CABLE LOCK

A cable lock is used to keep smaller pieces of equipment from being easy to steal. Many devices, most commonly portable computers, have a Kensington Security Slot, K-Slot, that is designed as a connection point for a cable lock. The K-Slot was originally developed by Kensington, which continues to develop new cable lock security devices.

SCREEN FILTERS

It may be worthwhile to install screen filters, also called privacy filters, which reduce the range of visibility of a screen down to a maximum of 30 degrees from perpendicular. These types of screens are designed to prevent someone sitting directly next to you, such as on an airplane, from being able to see the contents of your display.

VIDEO MONITORING

Video surveillance, video monitoring, closed-circuit television (CCTV), and security cameras are all means to deter unwanted activity and create a digital record of the occurrence of events. Cameras should be positioned to watch exit and entry points that allow access to any change in authorization level.

MOTION DETECTORS

A motion detector, or motion sensor, is a device that senses movement or sound in a specific area. Many types of motion detection exist, including infrared, heat, wave pattern, capacitance, photoelectric, and passive audio.

PHYSICAL ACCESS LOGS

Logs of physical access should be maintained. These can be created automatically through the use of smartcards for gaining access into the facility or manually by a security guide who will indicate entrance after inspecting each person's ID. The purpose of physical access logs is to establish context for logical logs produced by servers, workstations, and networking equipment. Infrared detection is often used by security cameras to see in perceived darkness or to detect movement in an area.

KEY & ASSET MANAGEMENT

Key management in relation to physical security focuses on the issuance of physical metal keys to those who need access into secured rooms, areas, or containers. A detailed log should be maintained of who was issued which key for which room/container and for what purpose. Regular auditing should be performed to ensure the responsible party still possesses the key and discloses whether or not the key has been exposed to theft or duplication. Keys should be numbered and, when possible, labeled or identified as not to be duplicated. Keys should be returned to the key manager when access is no longer required by that individual. When a worker who was in possession of a key is terminated, that lock and key should be replaced.

AN INTRODUCTION TO ACCESS MANAGEMENT

FEDERATION

Federation or federated identity is a means of linking a subject's accounts from several sites, services, or entities in a single account. Thus it is a means to accomplish single sign-on. Accountability logging is used to relate digital activities to humans. Federation creates authentication trusts between systems in order to facilitate single sign-on benefits. Federation trusts can be one-way or two-way and can be transitive or intransitive.

In a one-way trust, as when system A is trusted by system B, users from A can access resources in both A and B systems, but users from B can only access resources in B. In a two-way trust, such as between system A and system B, users from either side can access resources on both sides. If three systems are trust-linked using two-way nontransitive trusts, such as A links to B which links to C, then A resources are accessible by users from A and B, B resources are accessible by users from A, B, and C, and C resources are accessible by users from B and C. If three systems are trust-linked using two-way transitive trusts, then all users from all three systems can access resources from all three systems.

SINGLE SIGN-ON

Single sign-on means that once a user (or other subject) is authenticated into a realm, they need not re-authenticate to access resources on any realm entity. This allows users to access all the resources, data, applications, and servers they need to perform their work tasks with a single authentication procedure.

TRANSITIVE TRUST

Transitive trust or transitive authentication is a security concern when a block can be bypassed using a third party. A transitive trust is a linked relationship between entities where trust from one endpoint crosses over or through middle entities to reach the farthest linked endpoint.

A real-world example of transitive trust occurs when you order a pizza. The cook makes the pizza and passes it on to the assistant, who packages the pizza in a box. The assistant then hands the pizza to the delivery person, the delivery person brings it to your location to hand it to your roommate, and then your roommate brings the pizza into the kitchen, where you grab a slice to eat. Since you trust each link in the chain, you are experiencing transitive trust.

ACCESS CONTROL MODELS

Mandatory access control (MAC) is a form of access control commonly employed by government and military environments. MAC specifies that access is granted based on a set of rules rather than at the discretion of a user. The rules that govern MAC are hierarchical in nature and are often called sensitivity labels, security domains, or classifications. The main purpose of a MAC environment is to prevent disclosure: the violation of the security principle of confidentiality.

A government or military implementation of MAC typically includes the following five levels (in order from least sensitive to most sensitive):
- Unclassified
- Sensitive but unclassified
- Confidential
- Secret
- Top secret

DISCRETIONARY ACCESS CONTROL: DAC is the form of access control or authorization that is used in most commercial and home environments. DAC is user-directed or, more specifically, controlled by the owner and creators of the objects (the resources) in the environment. DAC is identity-based: access is granted or restricted by an object's owner based on user identity and on the discretion of the object owner. Thus, the owner or creator of an object can decide which users are granted or denied access to their object. To do this, DAC uses ACLs.

ACCESS CONTROL LIST: An ACL is a security logical mechanism attached to every object and resource in the environment. It defines which users are granted or denied the various types of access available based on the object type. Individual user accounts or user groups can be added to an object's ACL and granted or denied access.

ATTRIBUTE-BASED ACCESS CONTROL: ABAC is a mechanism for assigning access and privileges to resources through a scheme of attributes or characteristics. The attributes can be related to the user, the object, the system, the application, the network, the service, time of day, or even other subjective environmental concerns. ABAC access is then determined through a set of Boolean logic rules, similar to if-then programming statements, that relate who is making a request, what the object is, what type of access is being sought, and results the action would cause. ABAC is a dynamic, context-aware authorization scheme that can modify access based on risk profiles and changing environmental conditions.

ROLE-BASED ACCESS CONTROL: RBAC is another strict form of access control. It may be grouped with the non-discretionary access control methods along with MAC. The rules used for RBAC are basically job descriptions: users are assigned a specific role in an environment, and access to objects is granted based on the necessary work tasks of that role. RBAC is most suitable for environments with a high rate of employee turnover.

RULE-BASED ACCESS CONTROL: Rule-BAC is typically used in relation to network devices that filter traffic based on rules, as found on firewalls and routers. Rule-BAC systems enforce rules independent of the user or the resource, as the rules are the rules. If a firewall rule sets a port as closed, then it is closed regardless of who is attempting to access the system. These filtering rules are often called rules, rule sets, filter lists, tuples, or ACLs.

PHYSICAL ACCESS CONTROL

Physical access controls are needed to restrict physical access violations, whereas logical access controls are needed to restrict logical access violations.

PHYSICAL ACCESS CONTROLS: These should be implemented anywhere with a difference in value, risk, or use between one area of a facility and another; any place where it makes sense to have a locked door, technology-managed physical access controls should be implemented.

PROXIMITY DEVICES: A proximity device or proximity card can be passive, field-powered, or a transponder used to control physical access. The proximity device is worn or held by the authorized bearer. When it passes a proximity reader, the reader is able to determine who the bearer is and whether they have authorized access. A passive device reflects or otherwise alters the electromagnetic field generated by the reader. This alteration is detected by the reader. The passive device has no active electronics; it is just a small magnet with specific properties.

SMART CARDS: These are credit card–sized IDs, badges, or security passes with embedded integrated circuit chips. They can contain information about the authorized bearer that can be used for identification and/or authentication purposes. Some smart cards can even process information or store reasonable amounts of data in a memory chip. Many smart cards are used as the means of hardware-based removable media storage for digital certificates.

TOKENS

A token is a form of authentication factor that is something you have. It's usually a hardware device, but it can be implemented in software as a logical token.

An authentication token can be a hardware device that must be present each time the user attempts to log on. Often hardware tokens are designed to be small and attach to a keychain or lanyard. They are often referred to as keychain tokens or key fobs.

An authentication token can be a software solution, such as an app on a smart device. Since many of us carry a smartphone with us almost everywhere we go, having an app that provides OTP when necessary can eliminate the need for carrying around another hardware device or physical token. Software token apps are widely available and implemented on many Internet services; thus, they are easy to adopt for use as an authentication factor for a private network.

TIME-BASED ONE-TIME PASSWORD: TOTP tokens or synchronous dynamic password tokens are devices or applications that generate passwords at fixed time intervals. HMAC-based one-time password, HOTP, tokens or asynchronous dynamic password tokens are devices or applications that generate passwords based not on fixed time intervals but on a nonrepeating one-way function, such as a hash or HMAC operation.

IAAA

Identification and authentication are commonly used as a two-step process, but they're distinct activities. Identification is the assertion of an identity. This needs to occur only once per authentication or access process. Any one of the common authentication factors can be employed for identification. Once identification has been performed, the authentication process must take place. Authentication is the act of verifying or proving the claimed identity. The issue is both checking that such an identity exists in the known accounts of the secured environment and ensuring that the human claiming the identity is the correct, valid, and authorized human to use that specific identity.

Authorization is the mechanism that controls what a subject can and can't do, access, use, or view. Authorization is commonly called access control or access restriction. Most systems operate from a default authorization stance of deny by default or implicit deny. Then all needed access is granted by exception to individual subjects or groups of subjects. Once a subject is authenticated, its access must be authorized.

Accounting measures the resources a user consumes during access. This can include the amount of system time or the amount of data a user has sent and/or received during a session.

Accounting is carried out by logging of session statistics and usage information and is used for authorization control, billing, trend analysis, resource utilization, and capacity planning activities.

AN INTRODUCTION TO IDENTITY MANAGEMENT

MFA

Multifactor authentication is the requirement that a user must provide two or more authentication factors in order to prove their identity. There are three generally recognized categories of authentication factors.

- "Something You Are" is often known as biometrics.
  Examples include fingerprints, a retina scan, or voice recognition.
- "Something You Have" requires the use of a physical object.
- "Something You Know" involves information you can recall from memory.
  Examples include a password, code, PIN, combination, or secret phrase.
- "Somewhere You Are" is a location-based verification.
  Examples include a physical location or a logical address, such as a domain name, an IP address, or a MAC address.
- "Something You Do" involves some skill or action you can perform.
  Examples include solving a puzzle, a secret handshake, or a private knock. This concept can also include activities that are biometrically measured and semi-voluntary, such as your typing rhythm, patterns of system use, or mouse behaviors.

BIOMETRICS

Biometrics is the term used to describe the collection of physical attributes of the human body that can be used as an identification or authentication factor. Biometrics fall into the authentication factor category of something you are: you, as a human, have the element of identification as part of your physical body.

FINGERPRINT SCANNER: Used to analyze the visible patterns of skin ridges on the fingers and thumbs of people. Fingerprints are thought to be unique to an individual and have been used for decades in physical security for identification, and are now often used as an electronic authentication factor as well. Although fingerprint scanners are common and seemingly easy to use, they can sometimes be fooled by photos of fingerprints, black-powder and tape-lifted fingerprints, or gummy re-creations of fingerprints.

RETINAL SCANNERS: These focus on the pattern of blood vessels at the back of the eye. Retinal scans are the most accurate form of biometric authentication and are able to differentiate between identical twins. However, they are the least acceptable biometric scanning method for employees because they can reveal medical conditions, such as high blood pressure or pregnancy.

IRIS SCANNERS: These focus on the colored area around the pupil. They are the second most accurate form of biometric authentication. Iris scans are often recognized as having a longer useful authentication lifespan than other biometric factors because the iris remains relatively unchanged throughout a person's life.

VOICE RECOGNITION: This is a type of biometric authentication that relies on the characteristics of a person's speaking voice, known as a voiceprint. The user speaks a specific phrase, which is recorded by the authentication system. To authenticate, the user repeats the same phrase and it is compared to the original.

FACIAL RECOGNITION: Based on the geometric patterns of faces for detecting authorized individuals. Face scans are used to identify and authenticate people before accessing secure spaces, such as a secure vault. Many photo sites now include facial recognition, which can automatically recognize and tag individuals once they have been identified in other photos.

As with all forms of hardware, there are potential errors associated with biometric readers. Two specific error types are a concern: false rejection rate (FRR) or Type I errors and false acceptance rate (FAR) or Type II errors. The FRR is the number of failed authentications for valid subjects based on device sensitivity, whereas the FAR is the number of accepted invalid subjects based on device sensitivity.

The two error measurements of biometric devices, FRR and FAR, can be mapped on a graph comparing sensitivity level to rate of errors. The point on this graph where these two rates intersect is known as the crossover error rate, CER. FRR errors increases with sensitivity, whereas FAR errors decrease with an increase in sensitivity. The CER point is used to determine which biometric device for a specific body part from various vendors or of various models is the most accurate. The comparatively lowest CER point is the more accurate biometric device for the relevant body part.

IDENTIFY & ACCESS MANAGEMENT

IMPLEMENTATION OF ACCESS MANAGEMENT

FEDERATION

Federation is a way for multiple identity providers to verify a user identity. It uses standards and protocols like digital signatures, PKI, and encryption to map users across different service providers. Common examples are OAuth, OpenID Connect, and SAML.

SINGLE SIGN-ON

Single Sign-on is often used to describe a scenario where user credentials authentication not just one service provider, but many in a centralized system. Examples of Single Sign-On are Active Directory, LDAP, or utilize a central database/directory server.

TRANSITIVE TRUST

Transitive Trust is the idea that an organization only trusts an entity because it is trusted by an entity that already gained the organization's trust. For example, if a bank used a cloud service like AWS and that cloud service trusted Splunk, a tool to analyze big data, then transitive trust would be the bank making a deal with Splunk just because their partner, AWS, trusts them as a partner. Transitive Trust is common in banner ads and is significantly weaker than the other two forms of access control. A website will ask a trusted partner to maximize revenue and that partner will contract others and so on, unaware that some of the ad firms might be crimeware instead.

ACCESS CONTROL MODELS

Mandatory Access Control (MAC) is used in secure environments like financial institutions or military. A user is granted access based on their security clearance level. Each system and data is also protected not just by clearance level, but by sensitivity. Some systems or data are only granted on a "need to know" basis. MAC is either a confidentiality model or an integrity model. Confidentiality trusts that once a user is checked they are who they say they are and focuses on unauthorized access while Integrity checks the permissions of the user on each piece of data to prevent unauthorized modification.

DISCRETIONARY ACCESS CONTROL: DAC is the security on most personal desktop environments. The user who creates or owns an object can assign permissions on that object of whatever they want. If a user makes a folder, they can use their own discretion to decide what types of users can modify their file (outside of administrators). This is the only model that gives basic users the ability to handle permissions.

ATTRIBUTE-BASED ACCESS CONTROL: ABAC is a bit different from the other types; it uses policies to determine if access rights are granted to a user where policies can be a series of logic that can combine multiple attributes. For example, access control can be quite complicated, like requesting if the user is a manager of the department that the file is hosted from and getting a request back stating true or false and having a reaction based on that request, like granting write access. ABAC can handle requests in real time and in dynamic settings unlike all other types and is praised because of this.

ROLE-BASED ACCESS CONTROL: RBAC is similar to DAC, but users do not have discretionary access to permissions. A high level role like administrator makes access control dependent on a role and a user must be placed in that role to have permissions to the file. Everything is statically set and must be done by someone with an administrative-level. Forums often follow this model to help moderate chats and keep threads on topic, in guidance with the defined rules of the forum, and while making moderation fair.

RULE-BASED ACCESS CONTROL: Rule-BAC is when pre-defined rules mixed with mandatory and discretionary access control models is used to guide the system. For example, a rule saying that a type of user can only access files on a certain workstation at a certain time. It is different than ABAC because rules must be statically written and not request based. It is typically used for firewalls in an enterprise setting where the workstations and traffic matter for operations.

PHYSICAL ACCESS CONTROL

PROXIMITY CARDS: Have a microchip that has the sole purpose of providing the reader with the card's ID to be verified by a remote computer. These are commonly used today for door access to places and while proximity cards can have additional memory to make them "smart" it is typically not needed and typically only the ID numbers are read.

SMART CARDS: Can provide a lot of information including personal information, various security values, and a mode to authenticate. Smart cards have not only a microchip, but memory to store information. Smart cards can have multiple credentials to strengthen the handshake between the user and reader and can also be made for a purpose like a credit card or membership card.

TOKENS

HARDWARE TOKENS: Once inserted, these type in the OTP needed to authenticate a user.

SOFTWARE TOKENS: These provide an OTP to the user which the user can then enter into the software or hardware they wish to authenticate.

For Hardware and Software Tokens, there are two main ways to generate a password, HMAC-Based One Time Password (HOTP) and Time-Based One-Time Password (TOTP). HOTP requires a shared secret and a dynamic factor like a counter. A hash-based message authentication code from the other factor is generated and then the dynamic factor changes to modify the password each time. For TOTP, it works the same way, except instead of a physical moving factor it has a constantly changing one based on the time passed since epoch.

IAAA

Identification is something unique like an ID number, or name.

Authentication is some way to prove that the thing is the thing it is identified as (most likely with Multifactor Authentication covered below).

Authorization is using an Access Control Model, like the ones described above, to decide what the person is allowed to access.

Accounting is the act of tracing actions into logs to prove that a specific thing has a given action.

IMPLEMENTATION OF IDENTITY MANAGEMENT

MULTIFACTOR AUTHENTICATION (MFA)

Something you are is typically biometrics like fingerprint, iris scan, etc....

Something you have is typically a physical or digital device like the Tokens described above or a Passport (something only you could have obtained).

Something you know is a password or phrase you made to get into a said entity.

Somewhere you are is detecting location or device via IP or Mac Address.

Something you do is using a motion like a pattern or signature to verify that it is you.

BIOMETRICS

FINGERPRINT SCANNERS: Devices that grab a unique piece of data from finger patterns. They are commonly used today in smart phone devices as a quick way to securely authenticate the device and is in some newer laptops and used historically for physical security of buildings and rooms.

RETINAL SCANNERS: Much like fingerprint scanners, take a pattern of the body to create a unique piece of data used for authentication. Retinal scanners cast infrared light on a person's eye to map out a path on the retina while the person looks into the eyepiece. Like fingerprint scanning, retinal scanning must have close contract to get a good enough capture for use.

IRIS SCANNERS: Also use eyes for authentication, but instead of mapping out the retina, it uses camera technology to take a picture of the iris and then detects unique mathematical patterns in the iris. Iris scanning is fast, easy, and reliable since the mathematical patterns are unambiguous like finger and retinal paths. Iris scanning is very quick due to the unlikelihood of a false match and visibility of an iris.

VOICE RECOGNITION: Similar to the others in that it detects unique patterns within how the user talks to uniquely identify them.

FACIAL RECOGNITION: Instead of scanning just the iris with a camera, takes several shots of someone's face and uses those scans to predict if the person is who they say they are.

FALSE ACCEPTANCE RATE: FAR is the measure of how often an unauthorized user is identified as an authorized user by the biometric system.

FALSE REJECTION RATE: FRR is the measure of how often an authorized user is not identified as an authorized user by the biometric system.

CROSSOVER ERROR RATE: When the biometric system is tuned such that the smallest amount of error exists in which the other increases in rate; both FRR and FAR are at their smallest points.

INSTALLATION & CONFIGURATION OF IDENTIFY & ACCESS SERVICES

LIGHTWEIGHT DIRECTORY ACCESS PROTOCOL: LDAP is used to allow authenticated users to browse and locate objects in a distributed network database and is commonly used in Windows Active Directory (AD).

KERBEROS: An authentication protocol typically used in Active Directory. It is based on authentication tickets and timestamps for an authenticated user. Timestamps prevent relay attacks by making the time-to-live (TTL) very small for each ticket. Kerberos is made of several components. The first is Kerberos Key Distribution Center (KDC) which authenticates by giving out session keys and tickets. In active directory, the KDC is known as the Domain Controller. The next part is Authentication Service (AS) which verifies the user's identity using stored credentials in AD and then gives the user a Ticket-Granting Ticket (TGT) which is used to access all the resources the user needs on their domain. The next step in this process is called the Ticket-Granting Service (TGS) and is what makes TGT work. TGTs expire and need to be reissued by the AS periodically. When a user wants to access a resource, it presents its TGT to the TGS to authenticate and then gets a session key for communicating between the user and the resource server, known in Kerberos and AD as a service ticket, and is used for the duration of access

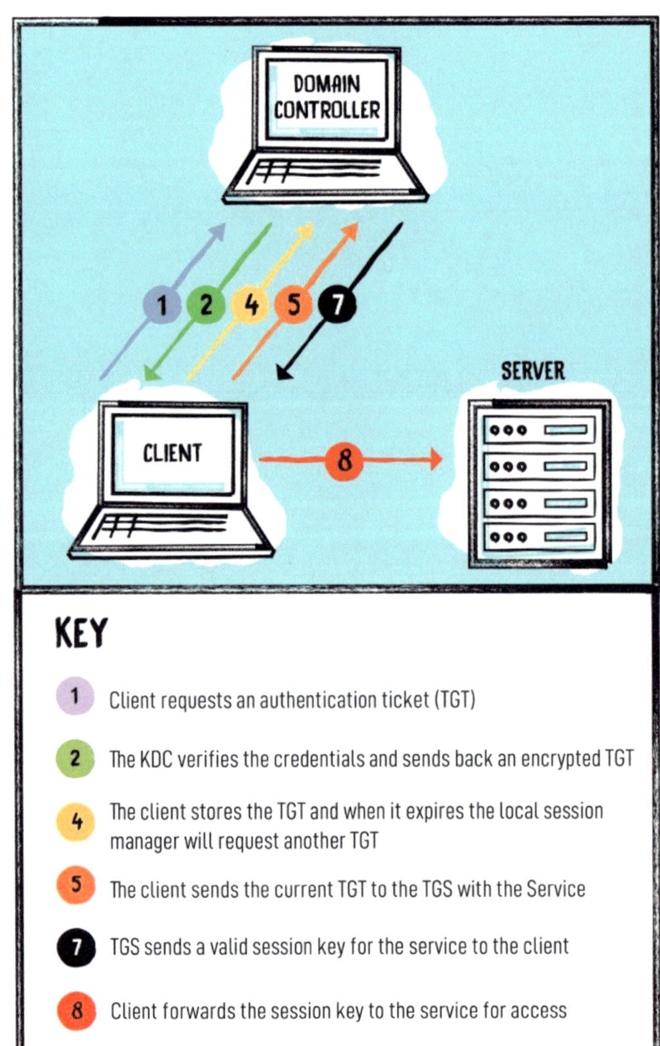

KEY

1 Client requests an authentication ticket (TGT)

2 The KDC verifies the credentials and sends back an encrypted TGT

4 The client stores the TGT and when it expires the local session manager will request another TGT

5 The client sends the current TGT to the TGS with the Service

7 TGS sends a valid session key for the service to the client

8 Client forwards the session key to the service for access

to the resource being requested. A new service ticket is needed for each request to the resource. While Kerberos is commonly used for Windows AD, it can be used in any operating system or LDAP-based implementation. A network that uses Kerberos to authenticate is known as a Kerberos realm. Kerberos uses port 88 and symmetric key cryptography.

TERMINAL ACCESS CONTROLLER ACCESS CONTROL SYSTEM PLUS: TACACS+ replaced RADIUS by allowing a remote user to connect and authenticate via a TACACS server. TACACS+ is a proprietary version of the protocol made by Cisco that enabled it to have newer and more secure authentication protocols like Kerberos and two-factor authentication. It later became an open standard. TACACS+ encrypts all traffic between all connection points so that it can include usernames and passwords and, because of this, is not backwards compatible with older versions of the protocol. TACACS and TACACS+ use port 49 by default.

CHALLENGE-HANDSHAKE AUTHENTICATION PROTOCOL: CHAP is an internet standard for authenticating users or a host to a centralized server. It is an older protocol that replaced sending username and password information over plaintext and is used primarily over PPP connections. CHAP works by having the authentication server send a challenge message to the user and hashing that with the user's password. Then, the authentication server will verify the challenge response by repeating the process with the known user's password. If the two strings match, then the user is known and authenticated.

PASSWORD AUTHENTICATION PROTOCOL: PAP is an older protocol that sends username and password data to a central authentication server. It was used for dial-up remote connection methods that are known to use Point-to-Point protocol. PAP transmitted information over plaintext, which is not considered secure since it can be easily intercepted by someone using a network sniffer like Wireshark.

MICROSOFT CHAP: MSCHAP was Microsoft's in-house CHAP used in early versions of Microsoft Windows. It has since been deprecated for more secure authentication protocols like Kerberos.

REMOTE AUTHENTICATION DIAL-IN USER SERVICE: RADIUS was developed to provide remote connections through dial-in services and provides Authentication, Authorization, and Accounting (AAA) services to clients and providers and is made up of many Request for Comments (RFCs). It uses ports 1812 for authentication and authorization and 1813 for accounting functions. A RADIUS client, unlike other protocols, is a network access server which remote hosts connect to. The RADIUS client acts as an intermediary that takes in a connection request and passes it along to the other servers, known as RADIUS servers. RADIUS servers can provide authentication servers for both UNIX-based systems and Windows. RADIUS is lacking in security. Communication between the RADIUS server and RADIUS client are encrypted, but communication from the RADIUS client and remote hosts is not and contains sensitive information like usernames and passwords which can be intercepted. RADIUS supports a variety of authentication methods including PAP and CHAP.

SECURITY ASSERTION MARKUP LANGUAGE: SAML is used to transfer information about authenticating users to an authentication service. It uses a XML-based standard to provide a format for passing information between services and can be used to pass identification and authorization information. SAML uses 3 types of information. The principle is what it calls the authenticating server. The identity provider is the part that authenticates the user. And the service provider is the part that accepts authentication information and, if correct, provides access to the resource needed. SAML is commonly used in the web for single sign-on.

OPENID CONNECT: Similar in function to SAML, but provides a service, not just to enterprise, but for everyone. OpenID was made for consumer apps and services while keeping enterprise in scope. It is an integral part of OAuth and offered by big players like Facebook, Google, and Twitter. OpenID is used to communicate from the end user to the resource party. The resource party asks the user for their OpenID and, once received, identifies the user via XPDS and generates a key to associate with the user. The request is then made to authenticate with user credentials. The user then enters in their creds and gets a session with the resource party to access what is needed.

OAUTH: A protocol that provides resource authorization. OAuth was created by Google, Twitter, and other big technology companies. It typically uses OpenID Connect to handle Single Sign-On, and then OAuth determines what the user should have access to. It is very popular today for modern websites like Facebook, GitHub, Google Apps, and more.

SHIBBOLETH: Similar to other single sign-on systems (SSO) but is able to support SSO to services outside of an organization while maintaining privacy. Unlike federated SSO and SSO, Shibboleth offers an embedded discovery service solution that works with the service provider needed in order to display an identity provider selector that can be integrated in whatever site is using it. Shibboleth can also provide user attributes to the identity provider. Shibboleth also implements SAML to be able to do Single Sign-On with HTTP/POST, artifacts, and push profiles. It is included in a lot of services that use SAML for identity management.

SECURE TOKEN: Responsible for issuing, renewing, validating, and canceling security tokens. Tokens issued identify people with the tokens to any services that follow a WS-Trust standard. Secure tokens help solve the problem of authentication across stateless platforms. The typical process is a user requests access with credentials, then the secure token validates creds, then provides a signed token to the client, the client then stores that token and sends it with each request for the servers to verify and respond with data.

NT LAN MANAGER: (NTLM) was developed by Microsoft to be the successor to LANMAN and is backwards compatible with that authentication protocol. It was not very secure on launch and quickly needed a v2. Both versions of NTLM work similarly to CHAP and are challenge-response authentication protocol. They were used in a lot of Windows versions until it was deprecated for Kerberos with NTLM v2 used as a secondary for tasks like authenticating to a server with an IP address, authenticating to another AD forest with legacy trust, and some other Windows authentication processes.

VARIABLES OF COMMON ACCOUNT MANAGEMENT PRACTICES

ACCOUNT TYPES

USER ACCOUNT: Used to authenticate and store information pertaining to a user. For a system to manage permissions of many different people, separation via user accounts is necessary. Users can be given permission and access to whatever they create and whatever the privileged accounts deem necessary.

SHARED ACCOUNTS: Not usually ideal since user accounts are made to track individual activity, but some are used in context, like guest accounts. When a guest account is shared, it is sometimes known as a generic account and is typically made to have a set level of functionality like a guest account, but for multiple people. Generic accounts are used in industry for things like kiosks, general use computer labs, and more. It isn't usually important to be able to track user activity for generic accounts since they are more than likely single purpose and highly trafficked.

GUEST ACCOUNTS: Made to give someone who doesn't typically use a workstation temporary access. Guests usually cannot create passwords, install software, or modify settings. Guests typically only have access to already installed applications like web browsers.

SERVICE ACCOUNT: Similar to a user account, but made for an application or service to interact with the OS, not a person. It is made to provide security context to a service while it is running. It can be assigned to user groups and given access and permission like with user accounts.

PRIVILEGED ACCOUNT: Made to give a user access to integral parts of a server. Privileged accounts can make meaningful changes and take full control of whatever processes and users have and are accessing. They have access to what is called the User Account Control (UAC) and are often referred to as administrators or root-level.

GENERAL CONCEPTS

LEAST PRIVILEGE: This means that an entity should only have the rights and privileges necessary to perform its task and nothing more.

ONBOARDING / OFFBOARDING: Refers to the process of adding someone to a project and removing someone from a project. During onboarding, proper account relationships should be initiated and the user should be put into the correct access control based on least privilege. During offboarding, the user should be removed from the access control groups and their account should be disabled.

PERMISSION AUDITING & REVIEW: An action that verifies that user accounts on a system are all needed and actually representing real people. It is important to do this often since users can come and go from groups and should not have retained permissions to a group they do not belong to.

USAGE AUDITING & REVIEW: Looking at logs to determine user activity. It is important to look at access control logs for root-level accounts due to their power and potential of misuse.

TIME-OF-DAY RESTRICTIONS: These help solve many account management problems. Since most employees have a set schedule, limiting permission to that account in off hours limits the amount of accounts to attack for attackers. It is really important to do this with privileged accounts due to the increased risk.

RECERTIFICATION: A process to re-verify user accounts. An example is running all the accounts against payroll to detect users that are no longer employed.

STANDARD NAMING CONVENTION: The idea of having a process for naming things. An example is naming servers with "development", "production", and "test" instead of something arbitrary so that users do not make an accidental change, or naming emails based on the first few letters of employees first name and last name so that people within the organization are easily contactable.

ACCOUNT MAINTENANCE: A routine screening where account attributes and permissions are checked and verified.

GROUP-BASED ACCESS CONTROL: The idea of managing access control based on groups instead of individuals. It is more efficient and also is less prone to error.

LOCATION-BASED POLICIES: These help with the idea of least privilege. A doctor should have access to clients in one clinic, but not in others owned by the same company. Location-based policies allow an organization to limit access and risk based on the location of a user or group.

ACCOUNT POLICY ENFORCEMENT

CREDENTIAL MANAGEMENT: The use of processes, software, and services to store, manage, and log the use of user credentials. This helps users manage their list of passwords, and some systems can provide a secure way to access the credentials across a wide range of platforms or in the cloud. Password managers, such as Lastpass or 1Password, are popular credential managers.

GROUP POLICY OBJECTS: These are used in Windows systems and act through a set of registry settings that are manageable via enterprise. There are a lot of settings that can be managed, including security settings, user credential settings, and more.

PASSWORD COMPLEXITY: The policy an organization puts in place to prevent passwords from being guessable. Typical requirements revolve around length and type of characters used.

EXPIRATION: When a user is no longer authorized to access an account. Having a manager take care of expiration is smart since they know when an employee quits, transfers departments, or no longer needs an account.

RECOVERY PLANS: These can seem pointless until a situation where you do not know the password occurs, or the person that has access isn't there. It is very important to have an alternative plan to gain access outside of credentials in case the credentials are not known.

DISABLEMENT: The go between for having access to an account and having the account disabled. When a user leaves a company, it is smarter to disable their account rather than remove it since removal can mess with permissions, ownership, and logs.

LOCKOUT: A mechanism to help prevent against brute force style attacks. If a user attempts to log into their account a set amount of times and fails, there can be a rule in place to lock the user out for a set amount of time.

PASSWORD HISTORY: A log of all previous passwords kept by the account. It is considered good security by some to not allow people to reuse passwords for a set number of passwords changes.

PASSWORD REUSE: This is considered a bad idea, and can lead to exposure of an account by someone who obtained a previously used password. A good rule of thumb is that passwords should not be reused for at least 1 year and 6 changes (whichever comes last).

PASSWORD LENGTH: This is a huge attribute to password strength. Password strength is determined by the entropy or randomness, and password length plays a huge part in increasing the entropy or randomness since they are based on the complexity and key space of a password.

RISK MANAGEMENT

POLICIES, PLANS, & PROCEDURES

STANDARD OPERATING PROCEDURE

Standard Operating Procedure is a defined method of performing some task or operation or a way of acting in a defined situation.

AGREEMENT TYPES

BUSINESS PARTNERS AGREEMENT: A BPA is a contract between two entities, dictating their business relationship. This can include things like a person's responsibilities, as well as the revenue, system, and data-sharing details. For example, in my training company, I entered into a Business Partnership with another company to produce an online training course on wireless hacking techniques. Our agreement clearly stated that I was responsible for writing the content and filming the videos, but the partner was responsible for providing a file server for me to upload the content, and they would then handle all of the editing and marketing.

SERVICE LEVEL AGREEMENT: An SLA is a contract between a supplier and a customer. SLAs are important because they set the tone for the relationship between the parties and will govern if and when things break down.

INTERCONNECTION SECURITY AGREEMENT: An ISA and is a formal declaration of the security stance, risks, and technical requirements of a link between two organizations' IT infrastructures.

MEMORANDUM OF UNDERSTANDING: MOU is an expression of agreement or aligned intent, will, or purpose between two entities.

PERSONNEL MANAGEMENT

MANDATORY VACATIONS: One or two weeks of mandatory vacation can be used to audit and verify the work tasks and privileges of employees. This often results in the detection of abuse, fraud, and / or negligence.

JOB ROTATION: This serves two functions: It provides a type of knowledge redundancy, and moving personnel around reduces the risk of fraud. data modification, theft, sabotage, and misuse of information.

SEPARATION OF DUTIES: The division of admin or privileged tasks into distinct groupings, with each group in turn assigned to unique admins. The application of separation of duties results in no single user having complete access to or power over an entire network, server, or system.

CLEAN-DESK POLICY: Used to instruct workers how and why to clean off their desks at the end of each work period.

BACKGROUND CHECKS: Used to verify that a worker is qualified for a position.

EXIT INTERVIEW: A controlled, respectful process of employee termination. The goal of an exit interview is to control the often emotionally charged event of a termination in order to minimize property damage, information leakage, or other unfortunate or embarrassing occurrences

ROLE-BASED AWARENESS TRAINING

DATA OWNER: Has full control over the objects that they own. This grants them the ability to make changes as well as set authorization for others.

SYSTEMS ADMINISTRATOR: The individual tasked with the responsibilities of implementing the security and functionality policies and requirements established by the organization and the system owner. Job functionality includes: installing updates, settings configuration, installing software, backing up data and settings, performing system evaluations, and responding to any troubleshooting or breach related issues.

SYSTEM OWNER: The entity responsible for setting the requirements for a system. They may be the organization as a whole or an individual network or IT manager. The System Owner crafts security policies, sets the baseline, and defines the configuration requirements of the system. Many times the System Owner and System Administrator are a combined job.

USER / OPERATOR: The individual who uses a deployed computer system to perform assigned day-to-day work tasks. The user has the responsibilities to accomplish their assigned work activities, perform those actions within the confines of the security policy and report any suspicious or abnormal events to the security staff.

PRIVILEGED USER: Someone granted additional capabilities, permissions, privileges, and user rights beyond those of a typical standard user. A privileged user is often referred to as an admin. A Privileged User or Admin may have special abilities over an area of the environment, such as database or firewalls, but should not have special abilities over all areas of the environment.

EXECUTIVE USER: Someone in a corporate management position who is granted additional privileges and capabilities beyond those of a typical standard user, but likely less than an admin or privileged user, especially if the person is not well versed in IT or security.

NON-DISCLOSURE AGREEMENT: An NDA is a contract that prohibits specific confidential, secret, proprietary, and/or personal information from being shared or distributed outside of a specific prescribed set of individuals or organizations

ONBOARDING: The process of adding new employees to the identity and access management system of the company. The onboarding process is used when an employee's role or position changes or when that person is awarded additional levels of privilege.

CONTINUING EDUCATION: Should primarily focus on improving efficiency, productivity, and security compliance. This also produces the side effect of employees gaining skills to move up inside the organization.

ACCEPTABLE USE POLICY & RULES OF BEHAVIOR: Define what is and is not acceptable activity, practice, or use for company equipment and resources. The acceptable use policy is specifically designed to assign security roles within the organization as well as prescribe the responsibilities tied to those roles. The policies define a level of acceptable performance and expectation of behavior and activity. Failure to comply with the policy can result in warnings and eventual termination.

ADVERSE ACTIONS: Consequences of failing to abide by a company's policies or actively committing criminal violations within the organization. This can be lectures from management, retraining or even criminal or civil prosecution.

GENERAL SECURITY POLICIES

When attempting to use Public Social Media networks/applications as an interface to customers, clients, and the public, be cautious and prepared when a message may get lost in the noise. If it is important to the organization, then you will have to brave the risks of social media use or host your own online social media tech. Self-hosted services can include discussion forums, text chats, and video conferencing. It is more safe for the organization have control over their social medium, then they can tamper down unwanted counter-messages.

Many organizations have a security policy regarding use of personal email while on company equipment. When a worker uses their personal email, then it exposes the system to malicious attachments, scripting in message bodies, and access to malicious and inappropriate content via hyperlinks. If the worker wishes to use a personal email, then allow them to use it through their personal device in order to limit company exposure.

BUSINESS IMPACT & RISK MANAGEMENT PROCESSES & CONCEPTS

RTO / RPO

RECOVERY TIME OBJECTIVE: RTO is the amount of time in which you think you can feasible recover the function in the event of a disruption. The goal of the business continuity planning, BCP, is to ensure your RTOs are less than your maximum tolerable downtime, MTD, resulting in a situation in which a function should never be unavailable beyond your MTD.

RECOVERY POINT OBJECTIVE: RPO is a measurement of how much loss can be accepted by the organization when a disaster occurs. This acceptable loss is measured in time. RPO is independent of RTO. Generally, backup systems are designed to prevent data loss over the RPO limit, and recovery solutions are designed to return things to normal before the RTO is exceeded.

MTBF

Mean Time Between Failures, MTBF, estimates established for each device or on prevailing best organizational practices for managing the hardware life cycle. MTBF is the expected typical time lapse between failures, such as the first failure and second failure.

MTTR

Mean Time To Repair / Restore, or MTTR, is the expected typical functional lifetime of the device, given a specific operating environment. MTTR is the average length of time required to perform a repair on a device. A device can go through numerous repairs before a catastrophic failure.

MISSION-ESSENTIAL FUNCTIONS

Mission-Essential Functions are any core business tasks that are central to the operation of the organization. These are the functions, processes, or tasks that if interrupted or terminated may cause the overall failure of the entire organization. Some organizations can survive for a brief period of time without functional mission-critical processes, but the maximum tolerable downtime, MTD, of these situations is often quite short.

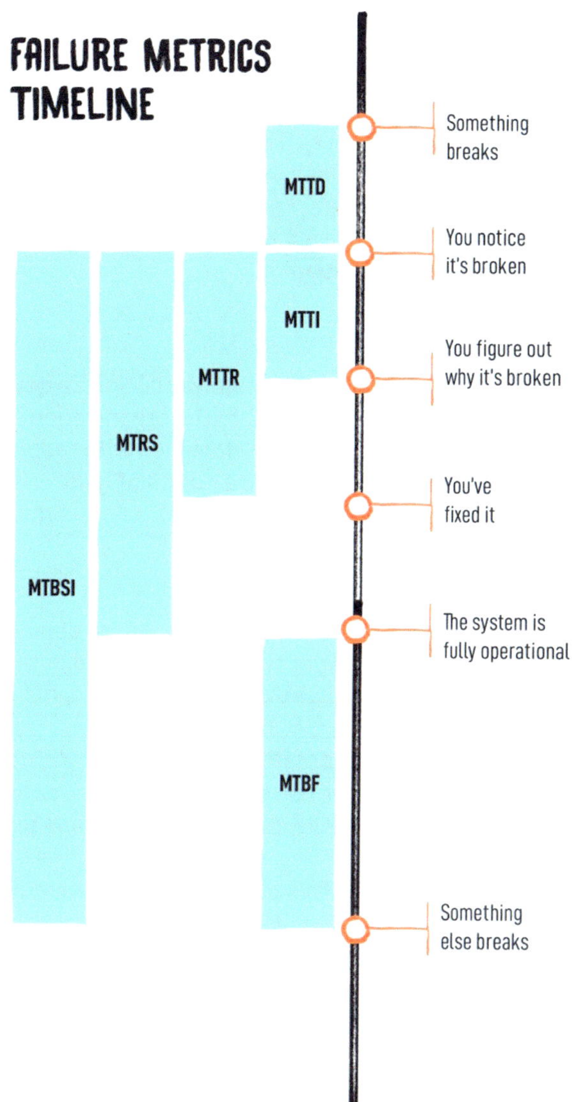

IDENTIFICATION OF CRITICAL SYSTEMS

Identification of critical systems is the process of evaluating risk and determining the best response to it. The critical elements or core components of an organization must be identified for long term survivability. The most critical systems are usually identified via the BIA process, which is the same process as risk assessment, except the BIA process focuses on business tasks, rather than assets. For each business task, an Annualized Loss Expectancy, ALE, is calculated. The processes, systems, or components that have the largest ALE are the most critical to the organization.

SINGLE POINT OF FAILURE

A Single Point of Failure is any individual or sole device, connection, or pathway that is moderately to mission-critically important to the organization. If one item fails, the whole organization suffers loss. Infrastructures should be designed with redundancies of all moderately or highly important elements in order to avoid single points of failure. Removes a Single Point of Failure, one must add redundancy, recovery options, or an alternative means to perform busin4ess tasks and processes.

IMPACT

Impact is the measurement of the amount of damage or loss that could be or would be caused if a potential threat is ever realized. This is indicated by the Exposure Factor, EF, which is the percentage of asset value that would occur if a risk has happened.

LOSS OF LIFE: It is imperative to implement protections to reduce or even eliminate the potential risks of loss of life. This can be tornadoes, fire, active shooter, etc. Most companies have policies in place, such as fire drills, tornado drills, escape plans, etc.

PROPERTY DAMAGE: This can potentially impact of a wide range of physical security breaches, but might also be a result of social engineering and logical/technical attacks. Property includes facility, equipment, and utilities.

SAFETY: Relating to the well-being of personnel. Safety violation defenses should include protections against physical and physiological harm to individuals. Organizations are required to provide a safe work environment, such as proper lighting, stable flooring, handrails, marking where vehicles are allowed to travel, and so on.

SECURITY BREACH: This can have an impact on the financial future of the organization. Recovering from an intrusion can be expensive, especially if hardware must be replaced, if lawsuits are filed, or if an investigation is required to collect evidence. Some organizations may purchase insurance against such dangers, but the organization must calculate the cost effectiveness of insurance and whether it covers any issues that might actually be experienced.

A security breach can damage an organization's reputation. If a security breach becomes public knowledge, then media outlets might emphasize the negative aspects of the intrusion rather than any successful detection, response, and recovery or preventative measures put into place postfix.

PRIVACY IMPACT ASSESSMENT

A Privacy Impact Assessment, or PIA, is used to determine privacy risks, how to mitigate those risks and whether to notify the affected parties as related to a new future project or endeavors. A PIA should be drafted when an organization will be collecting PII from its employees and/or its customers through a new software application, outreach program, or any other type of real-world or digital interaction. The PIA should then evaluate the means of collection, storage, protection, use, distribution, access, and sharing of the PII.

PRIVACY THRESHOLD ASSESSMENT

A Privacy Threshold Assessment, or PTA, is used to evaluate the data that an organization has already collected to determine whether such data is PII, business confidential, or non-sensitive data. If a PII is discovered, then its source must be determined and its intended uses uncovered. If necessary, additional safeguards to protect against PII distribution are to be implemented. The PTA can then be used to determine whether the PII is illegitimate and should be purged and the related parties notified.

THREAT ASSESSMENT

Strong organizations have plans and procedures in place to help mitigate the effects a disaster has on their continuing operations and to speed the return to normal operations.

ENVIRONMENTAL THREATS: Natural disasters that are triggered by forces outside of human control, such as tornadoes or earthquakes. These threats can result in anything from minor interruptions to tasks, to complete organizational collapse.

MAN-MADE THREATS: Any event or occurrence caused by humans. This can range from acts of war and terrorism to data theft, sabotage, and embezzlement.

Organizations need to treat any and all threats seriously. They need to understand that there are significant security risks that originate both internally and externally. Outside attackers are often given more of a spotlight than internal attackers, but internal attackers can cause more significant or lasting damage to an organization.

RISK ASSESSMENT

SINGLE LOSS EXPECTANCY: SLE is the potential dollar value loss from a single risk-realization incident. To obtain this metric, multiply the Exposure Factor by the asset value. Exposure Factor is calculated by using historical data from the previous occurrences in order to predict the amount or percentage of loss that might occur when and if the threat causes harm in the future.

ANNUALIZED LOSS EXPECTANCY: ALE is the potential dollar value loss per year per risk. Its calculated by multiplying the Single Loss Expectancy by the Annual Rate of Occurrence. Once the ALE is calculated, they are put in a list from greatest to least. This list should then guide security policy designs, starting from the top of the list.

ANNUAL RATE OF OCCURRENCE: ARO is the probability that a specific risk may occur a number of times each year. This Annual Rate of Occurrence is obtained by insurance companies, historical records, or by guessing.

ASSET VALUE: AV is the worth of an asset to an organization, found by calculating the tangible and intangible value, expense and costs. AV predicts the amount of loss the organization would suffer if the asset was no longer available or viable.

RISK REGISTER: A document that inventories all of the identified risks to an organization or system within an individual project. A risk register is used to record and track the activities of risk management, including: identifying risks; evaluating the severity and priority of those risks; prescribing responses to reduce or eliminate the risks; tracking the progress of risk mitigation.

LIKELIHOOD OF OCCURRENCE: The measurement of probability that a threat will become realized within a specific time period. Within the scope of risk assessment, likelihood is measured on a yearly basis. This measurement is also called the ARO.

When evaluating organizational risk, it is important to consider external factors that can affect the organization, especially related to company stability and resource availability. This is called a Supply Chain Assessment, in which an organization's supply chain is analyzed to determine what risks exist. Is the organization operating on a just-in-time basis where materials are delivered just before or just as they are needed? What if there are delays in delivery? Are there any surplus or buffer materials that can be used to supplement manufacturing?

The goal of risk management is to minimize Impact. This can be accomplished through prevention of risk, and realization of quick, sufficient response to harm.

QUANTITATIVE RISK ANALYSIS: Assigns real dollars to the loss of an asset. Quantitative risk analysis results in probability percentages. It creates a report that has dollar values for level of risk, potential loss, cost of countermeasures, and value of safeguards. This is not a complete analysis and must be paired with Qualitative Risk Analysis.

Qualitative Risk Analysis assigns subjective and intangible values to the loss of an asset. It is more scenario based than calculator based. The analyzer ranks threats on a scale to evaluate risk, cost, and effects. A qualitative risk matrix is used to show the relationships between frequency and consequence. This method of risk analysis requires sound judgement, intuition and experience.

TESTING

Before performing criminal violation simulations, it is important to obtain Penetration Test Authorization from senior management and scope the penetration test. Any unauthorized activity out of scope is a violation of the company's security policy and potentially criminal activity.

Authorization to perform vulnerability tests should be obtained before any security assessment. Because of the potential risk of data loss or system crashes, it's often prohibited for typical users. It could also lead to the discovery of a critical vulnerability that could cause serious harm to the organization. Due to this risk, most organizations don't allow unauthorized vulnerability testing.

RISK RESPONSE TECHNIQUES: Essential techniques to a successful Risk Analysis:
- A complete and detailed evaluation of all assets
- A complete list of all threats / risks, rate of occurrences, and extent of losses if they occur.
- List threat specific safeguards / countermeasures and identify effectiveness and ALEs.
- A cost-benefit analysis of each safeguard.

RESPONSES TO POTENTIAL RISKS:
- Accept or Tolerate
- Assign or Transfer
- Avoid
- Reduce or Mitigate
- Reject or Ignore

Accepting Risk or Tolerating Risk is a valuation by management of the cost-benefit analysis of possible safeguards and the determination that the cost of the countermeasure greatly outweighs the possible cost of loss due to a risk. It means management has agreed to accept the consequences and the loss if the risk is realized.

Assigning Risk or Transferring Risk is placing the cost of loss that a risk represents onto another entity or organization. This is done through purchasing insurance or outsourcing

A variation of assigning risk is Avoiding Risk. This is the process of selecting alternate options of activities that have less associated risk than the default, common, expedient, or cheaper option. This can also be the process of adding deterrents to risk, would-be violations of security and policies. This could be adding multi-factor authentication or putting security cameras outside.

Reducing Risk or Mitigating Risk is the implementation of safeguards and counter measures to eliminate vulnerabilities or block threats. Picking the most cost-effective or beneficial countermeasure is part of risk management.

CHANGE MANAGEMENT

The main goal of change management is to prevent change from causing unnecessary downtime or reductions in security. This is an important element of any risk management strategy.

INCIDENT RESPONSE & FORENSICS

INCIDENT RESPONSE PLAN

Every organization needs an incident response plan, IRP. The IRP is the SOP (Standard Operating Procedure) that defines how to prevent incidents, how to detect incidents, how to respond to incidents, and how to return to normal when the incident is concluded.

An incident occurs when an attack, or other violation of your security policy, is carried out against your system. There are many ways to classify incidents; here is a general list of categories:

- Scanning
- Data breach
- Malicious code
- Denial of service

The IRP should detail the Roles and Responsibilities of each member of the incident response team. The overall responsibilities include prevention, detection, and response.

After an incident has been contained, the incident response team is responsible for fully documenting the incident and making recommendations about how to improve the environment to prevent a recurrence.

Although it is not required, many organizations now have a dedicated team responsible for investigating any computer security incidents that take place. These teams are commonly known as cyber incident response teams (CIRTs) or computer security incident response teams (CSIRTs). When an incident occurs, the response team has four primary responsibilities:

- Determine the amount and scope of damage caused by the incident.
- Determine whether any confidential information was compromised during the incident.
- Implement any necessary recovery procedures to restore security and recover from incident-related damages.
- Supervise the implementation of any additional security measures necessary to improve security and prevent recurrence of the incident.

The members of the CIRT should be involved in continuing education and research to keep their knowledge current on new attacks and exploits as well as responses and repair options. But this knowledge needs to be tested in a real-world simulation prior to use in an actual incident event.

INCIDENT RESPONSE PROCESS

An incident response procedure is to be followed when a security breach or security violation has occurred. One of the most important goals of incident response is containment: the protection and preservation of evidence.

Preparation is necessary to ensure a successful outcome of unplanned downtime, security breaches, or disasters. Being prepared includes defining a procedure to follow in response to incidents, buttressing an environment against incidents, and improving detection methods.

The first step in responding to an incident is to detect and become aware that an incident is occurring. Without detection, incidents would instead be false negatives (the lack of an alarm in the presence of malicious activity)—in other words, unknown unknowns. If an organization is not aware that it is actively being harmed, then it doesn't know there is a need to respond or make changes. Detection can come from anywhere inside or outside the organization, whether via automated systems (such as IDSs and internal employees) or external groups (such as the FBI or Interpol). Regardless of source, the CIRT is the team responsible for responding.

Containment means to limit the scope of damage and prevent other systems or resources from being negatively affected. Containment is especially important when the incident includes virus infection, remote control access, a Trojan horse, a logic bomb, or the use of hacker tools. Malicious use of these components may leave residual elements that are activated at a later time. Thus, after containment, eradication of malware should be part of the recovery process in order to prevent further damage or loss.

Eradication includes the processes used to remove or eliminate the causes of the incident, such as removing software, deleting malware, changing configurations, firing personnel, disabling compromised accounts, and blocking IP addresses and ports. In some instances, eradication is not necessary or is handled in the recovery phase. In some cases, the act of restoring systems from backups performs the eradication of the offending application or malicious software.

Recovery is the process of removing any damaged elements from the environment and replacing them. This can apply to corrupted data being restored from backup and to malfunctioning hardware or software being replaced with updated or new versions.

A final step in incident response is to evaluate the response plan and procedures and improve them as necessary. This review can also serve as a means to extract or clarify lessons learned during an incident response.

ORDER OF VOLATILITY

Volatility is the likelihood that data will be changed or lost due to the normal operations of a computer system and the passing of time. Collection of potential evidence should be prioritized based on the type of event, incident, or crime as well as the order of volatility. For example, RAM is wiped when a computer is turned off, so it should be a higher collection priority than system logs, which are saved to disk and can be retrieved after a system is rebooted.

CHAIN OF CUSTODY

Chain of Custody is a document that indicates various details about evidence across its life cycle. It begins with the time and place of discovery and identifies who discovered the evidence, who secured it, who collected it, who transported it, who protected it while in storage, and who analyzed it. Ultimately, the chain-of-custody document details all persons who had controlling authority over and access to the evidence.

LEGAL HOLD

A legal hold is an early step in the evidence collection or e-discovery process. It is a legal notice to a data custodian that that specific data or information must be preserved and that good-faith efforts must be engaged to preserve the indicated evidence.

DATA ACQUISITION

Data acquisition is the processes and procedures by which data relevant to a criminal action is discovered and collected. In most situations, evidence collection should be performed by licensed and trained forensics specialists, usually those associated with law enforcement.

During a criminal investigation, capturing the System Image is the process of creating an image backup. The forensic duplication system calculates a hash of the original media before and after the bitstream image copy is performed.

In some network environments, it may be possible to maintain an ongoing recording of network traffic and log them. However, because this would result in a massive need for storage capacity, such a recording only maintains a sliding window of recent network activity—often measured in minutes or, at most, hours. If a violation is detected promptly, the window of network traffic can be preserved for more detailed offline analysis.

There are two issues related to video. First, if security cameras are present and video was captured of a security violation, those captured video images need to be preserved as evidence. Video (and audio) recordings may also track sensitive data that has been input, such as credit card numbers. Second, while performing an investigation, especially while seeking out physical and/or logical evidence, it can be important to have someone videotape the process. The videotaped observation can assist in crime scene reenactments, orientation, and the proper explanation of evidence during a presentation in court.

As an event is recorded into a log file, it is encoded with a time stamp. The time stamp is pulled from the clock on the local device where the log file is written or sent with the event from the originating device if remote logging is performed.

As mentioned in the "Capture System Image" section earlier, it is important to take a hash of a storage device before and after image duplication.

When performing a forensic investigation, never trust the software on the suspect's computer. Thus, using native screen-capture tools or features is not recommended. Instead, use a camera to take photographs of anything being displayed.

A witness is someone who experienced an event or incident through one or more of their five senses. A witness can provide information about what occurred, where the occurrence took place, and the chronological order of related events.

PRESERVATION

Forensic preservation aims at preventing any change from occurring as related to collected evidence. These efforts include removing relevant storage devices from their systems, using write-blocking adapters to block any writing signals from being received by storage devices, using hash calculations before and after every operation, and analyzing only cloned copies of storage devices and never the original device.

RECOVERY

During forensic investigations there is often a need to recover or restore data in order to make it usable and to determine whether it is related to the criminal activity. One means of recovery is to restore files from backup, whether those backups are hard drives, tapes, optical discs, or the cloud.

STRATEGIC INTELLIGENCE / COUNTERINTELLIGENCE GATHERING

Strategic intelligence gathering requires investigative and interviewing skills, which some law enforcement officers, military, and deputized civilians use to discover information that may be relevant to a criminal activity.

Active Logging is used to gather and maintain a wide range of security- and system-related events. Active logging may be a more thorough collection of events as well as a more robust means of preserving and retaining the audit trails for the purpose of forensic investigation rather than just overall system management and uptime optimization.

TRACK MAN HOURS

Throughout the implementation of an incident response procedure or forensic investigation, you should document every action taken by end users and the incident-response/investigative teams. This documentation will serve as an audit trail to retrace the actions taken and the events that occurred during the incident.

PRIVACY PRACTICE, CONTROLS, & CONTINUITY OF OPERATIONS

RECOVERY SITES

HOT SITE: A real-time, moment-to-moment mirror image of the original site. It contains a complete network environment that is fully installed and configured with live current business data. The moment the original site becomes inoperable due to a disaster, the hot site can be used to continue business operations without a moment of downtime. Hot sites are the most expensive type, but they offer the least amount of downtime.

WARM SITE: A partially configured alternate site with most of the server and networking infrastructure installed. In the event of a disaster, some final software installation and configuration are needed, and data must be restored from a backup set. A warm site may require hours or a day to get it ready for real-time operation to support the business's mission-critical functions. A warm site is moderately costly, but it is a realistic option for recovery if the organization can survive a few days of downtime.

A COLD SITE: Often little more than an empty room. It can be a location with no equipment or communications at all, or it can be a site with equipment in boxes and essential communications and utilities connected. In either case, it may require weeks of work to set up and configure in order to support the company's processing needs.

ORDER OF RESTORATION

Order of Restoration is the order in which a recovery effort should proceed. In most situations, when a disaster strikes, the most mission-critical business processes should be restored first. Then, other processes in descending criticality should be repaired.

BACKUP CONCEPTS

Backups are an essential part of business continuity because they provide insurance against loss of data files. The mantra of all security professionals should be: backup, backup, backup.

DIFFERENTIAL BACKUP: Copies only those files with a set or flagged archive bit. It doesn't alter the archive bit, thus selecting only those files that are new or that have changed.

INCREMENTAL BACKUP: Copies only those files with a set or flagged archive bit. It clears or resets the archive bit, thus selecting only those files that are new or that have changed.

SNAPSHOTS: Typically related to virtual machines, VMs, where the hypervisor is able to make a live copy of the active guest OS. Snapshots are complete copies of a VM that might take only a few minutes to create compared to hours for cloning hard drives, which typically must be done offline.

FULL BACKUP: Copies all files to the backup media regardless of the archive bit setting. It clears or resets the archive bit.

> **NOTE:** Incremental and differential backups are performed in concert with full backups. For example, a full backup could be performed at the beginning of each week, and then daily incremental or differential backups could be performed the other six days. Daily incremental backups consume approximately the same amount of time and storage space each day, whereas differentials grow larger and take longer each day.

GEOGRAPHIC CONSIDERATIONS

When drafting business continuity and disaster recovery plans, you should assess the geographic considerations and their impact on preparation and responses.

A backup strategy should include storage of backup media offsite to provide a reliable means of restoration and recovery in case of a significant event that may harm primary business operations. Offsite backups are a reliable means of recovery due to major damage to the primary production environment.

Alternate processing facilities and offsite backup storage should be a reasonable distance away from the primary site. What is reasonable is subjective, but it depends on the value of the assets and the risk to an organization. Generally, alternate facilities should be far enough away that they will not be affected by the same disaster that harms the primary location, but not so far away that it is overly inconvenient to travel to the alternate facility while the primary location is repaired.

Location selection should include consideration of accessibility, such as the number and size of roads leading to the facility, utility access and reliability, local crime rate, local hazards, and the likelihood of natural disasters affecting the area.

Some industries and individual organizations may be bound or limited by laws and regulations as to where their primary processing facilities can be located. This may be a limitation to stay within a country's borders or to stay away from populated areas.

Data sovereignty is the concept that, once information has been converted into a binary form and stored as digital files, it is subject to the laws of the country in which the storage device resides.

CONTINUITY OF OPERATIONS PLANNING

A tabletop exercise is a discussion meeting focused on a potential emergency event. It is usually performed verbally or with minimal visual aids, such as blueprints, charts, or board game miniatures representing resources. It is a means to walk through and evaluate an emergency plan in a stress-free environment.

All teams should facilitate an After-Action Report, AAR, of the incident within a week of the occurrence to ensure that key players in the incident share their knowledge and develop best practices to assist in future incident response efforts. The AAR should detail the issues involved, the responses attempted, the successful resolution, and any oversights, mistakes, or lessons learned. These reports are useful in preparing for future incidents, defending the organization in court, and training future members of response teams.

The use of redundant servers is another example of avoiding single points of failure. A redundant server is a mirror or duplicate of a primary server that receives all data changes immediately after they are made on the primary server.

Alternate business practices are any secondary, backup, fail-back, or fallback plans that can be used in the event that the preferred recovery strategies and planning procedures fail. A backup contingency plan is an alternate solution or response in case the primary plan fails or is not as successful as planned.

An Alternate Processing Site is a secondary location where the business can move and continue performing mission-critical business operations. There are three levels of alternate sites: hot, warm, and cold.

DETERRENT

A Deterrent access control is deployed to discourage violation of security policies. Deterrent and preventive controls are similar, but deterrent controls often depend on individuals deciding not to take an unwanted action.

PREVENTIVE

A Preventive access control is deployed to thwart or stop unwanted or unauthorized activity from occurring. Examples of preventive access controls include fences, locks, biometrics, mantraps, lighting, and alarm systems.

DETECTIVE

A Detective access control is deployed to discover or detect unwanted or unauthorized activity. Detective controls operate after the fact and can discover the activity only after it has occurred. Examples of detective access controls include security guards and motion detections.

CORRECTIVE

A Corrective access control modifies the environment to return systems to normal after an unwanted or unauthorized activity has occurred. It attempts to correct any problems that occurred as a result of a security incident. Corrective controls can be simple, such as terminating malicious activity or rebooting a system.

COMPENSATING

A Compensating access control is deployed to provide various options to other existing controls to aid in enforcement and support of security policies. It can be any control used in addition to, or in place of, another. For example, an organization may dictate that all PII must be encrypted.

TECHNICAL

Technical access involves the hardware or software mechanisms used to manage access and to provide protection for resources and systems. As the name implies, it uses technology. Examples of Technical access controls include authentication methods, such as usernames, passwords, smart cards, and biometrics.

ADMINISTRATIVE

Administrative access controls are the policies and procedures defined by an organization's security policy and other regulations or requirements. They are sometimes referred to as management controls. These controls focus on personnel and business practices. Examples of Administrative access controls include policies, procedures and hiring practices.

PHYSICAL

Physical access controls are items you can physically touch. They include physical mechanisms deployed to prevent, monitor, or detect direct contact with systems or areas within a facility.

DATA DESTRUCTION & MEDIA SANITIZATION

BURNING: Incineration can be an effective means of destroying paperwork as well as media storage devices. Care should be taken to address any toxic fumes that may be produced when burning hardware, and the leftover materials may be recycled.

SHREDDING: An effective destruction technique for both paperwork and media storage devices; however, different equipment will be needed for these two techniques. Paperwork shredding should generally be accomplished using a cross-cut shredder. Media device shredding is accomplished using an industrial metal shredder or chipper.

PULPING: A paperwork destruction process that involves shredding paper and mixing it with a liquid to create a fibrous mush.

PULVERIZING: A means of device destruction that goes beyond the shredding level to a point where the devices are reduced to a grain or powder. When devices are pulverized, it prohibits any data remnant recovery.

DEGAUSSING: A means of media storage device data destruction using strong magnetic fields. It is effective only on magnetic media, such as hard drives and tapes; it is not effective on other forms of media, such as optical discs, SSDs, and flash memory cards.

PURGING: Frequently a reference to the level or quality of a data destruction process rather than an actual procedure that should be followed.

DATA WIPING: The process of removing data from a storage device. Often the intention is to prevent data remnants from being recovered that would lead to data leakage.

DATA SENSITIVITY LABELING & HANDLING

HIGHLY CONFIDENTIAL: Generally relates to the most valuable and sensitive data level in a business classification scheme.

PRIVATE: Relates to individually related data (PII) in a business classification scheme.

PUBLIC: Relates to the least sensitive data level in a business classification scheme.

PROPRIETARY DATA: A form of confidential information. If Proprietary data is disclosed, it can have drastic effects on an organization's competitive edge.

PERSONALLY IDENTIFIABLE INFORMATION: PII is any data item that is linked back to the human from whom it was gleaned. PII that is medically related is protected under HIPAA laws. Companies should clearly disclose what PII is collected and how it will be used in the acceptable use policy.

PROTECTED HEALTH INFORMATION: PHI according to the laws of the United States, is any data that relates to the health status, use of health care, payment for health care, and other information collected about an individual in relation to their health. The U.S. Health Insurance Portability and Accountability Act (HIPAA) defines PHI in relation to 18 types of information that must be handled securely to protect against disclosure and misuse.

DATA ROLES

OWNER: An information owner is the person who has final corporate responsibility for classifying and labeling objects and protecting and storing data. The owner may be liable for negligence if they fail to perform due diligence in establishing and enforcing security policies to protect and sustain sensitive data.

CUSTODIAN: Also called a steward, this is a subject who has been assigned or delegated the day-to-day responsibility of properly storing and protecting objects.

PRIVACY OFFICER: A company executive tasked with the responsibilities of crafting the company privacy policy, implementing that policy, and overseeing its operation and management. The goal of the privacy officer is to ensure that personal data related to employees and customers is properly handled and protected.

DATA RETENTION

A retention policy defines what data is to be maintained and for what period of time. A retention policy defines the parameters and operations of data retention. Retention policies may also need to define the purpose of the held data, the security means implemented to protect the held data, and the officers of the organization who are authorized to access or handle the held data.

LEGAL AND COMPLIANCE

Every organization needs to verify that its operations and policies are legal and in compliance with their stated security policies, industry, obligations, and regulations.

AUDITING: This is necessary for compliance testing, also called compliance checking. Verification that a system complies with laws, regulations, baselines, guidelines, standards, best practices, and policies is an important part of maintaining security in any environment. Compliance testing ensures that all necessary and required elements of a security solution are properly deployed and functioning as expected.

COMPLIANCE CHECKS: These can take many forms, such as vulnerability scans and penetration testing. They can also use log analysis tools to determine whether any vulnerabilities for which countermeasures have been deployed have been attempted or exploited on the system.

CRYPTOGRAPHY & PKI

BASIC CONCEPTS OF CRYPTOGRAPHY

SYMMETRIC ALGORITHMS & STREAM VS BLOCK

A Symmetric Algorithm uses one key to encrypt and decrypt data, commonly known as a secret key, because in order for it to be secure, the key used needs to be private. The most popular type of symmetric or "secret key" algorithm is AES which we will discuss later on in the guide.

These one key encryption methods can be split up into two categories: stream and block. Stream Algorithms encrypt data byte by byte while Block Algorithms encrypt data block by block (where a block is multiple bytes defined by the program).

MODES OF OPERATION

Most Symmetric Algorithms support several Modes of Operations, including ECB, CBC, CTM, and GCM.

ASYMMETRIC ALGORITHMS

Another common type of encryption algorithm is Asymmetric. Asymmetric is when you use two keys to encrypt information, and is commonly known as "public key" encryption since you share one of your keys and keep the other private. One key decrypts the other. If Zach and Jake, want to share information and maintain secrecy or authentication, they can utilize Asymmetric encryption. Zach can encrypt a message with his private key and share it. Jake and all other people know Zach must have sent the message because only his public key decrypts it. Likewise, Zach can encrypt a message with Jake's public key and then only Jake can read the message since he is the only one with his private key. The first example is a way to authenticate that someone is who they say they are. The latter is a great way to keep secrecy and ensure only the intended recipient can read the message.

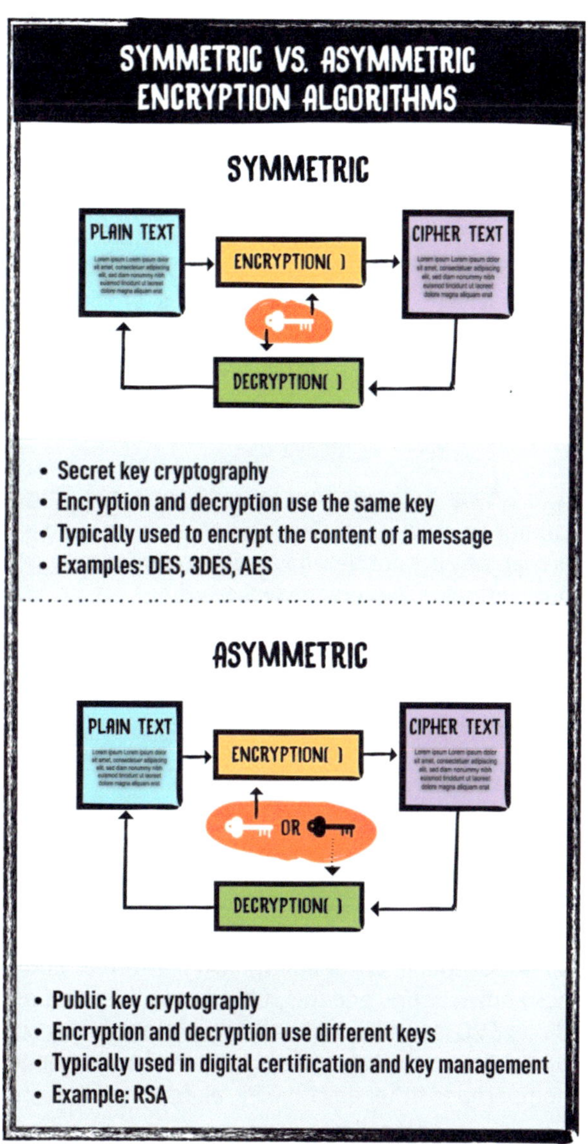

HASHING & COLLISION

Hashing is the idea of building unique keys for data. A Hashing Algorithm will take a large amount of data and convert it to a small practical value to use as the data's index.

Since hashing algorithms convert large amounts of data into small values, there is a possibility of it creating the same value for two sets of data. This is called a Collision.

SALT, IV, NONCE

Salts, Initialization Vectors (IV), and Nonces, are values in cryptography made to be used one-time, but are distinctly different. In a nutshell, salts and IVs are random and a nonce is iterative and all of them are typically assumed to be identifiable by attackers.

SALT: Random data that is placed under a certain context. The main purpose is to make dictionary style attacks not feasible. Basically, in a password-based system, if there is not a salt, an attacker could create a dictionary of common passwords and look up what a user's password is through the authenticator. Using a salt means the attacker would have to make a new dictionary for all salt possibilities which given the quantity, makes the attack not worth it. These can be generated on each user's OS, based on session keys, or some other way, but should be completely random.

NONCES: Data placed into different cryptographic protocols and algorithms. They can either be sequential or random. Sequential guarantees that nonces are not repeated, can take from algorithms that already have a sequential values to save space, and prevent relay attacks.

IVS: Similar to Nonces but are NOT the same. They follow the same structure as a nonce, but cannot be predictable. Meaning there is no such thing as a sequential IV. One downside to IVs is that they cannot defend against capture-replay attacks since they are random and it isn't guaranteed that everyone was used.

To recap, salts are added on to a hash to make them more unpredictable, nonces are sequential or random bits of data inputted into an algorithm (not after), and IVs are like nonces, but must be completely random.

ELLIPTIC CURVE

A type of cryptography that is one of the most powerful. Commonly referred to as ECC, it is used in SSL for data transfers within servers, on weaker mobile devices, and lots of other security systems. An Elliptic Curve is a set of points that satisfy a math equation and can typically be defined as $y^2 = x^3 + ax + b$. Elliptic Curves are used in cryptography by choosing a prime number as a max, a random curve equation in the format above, and a point on the curve. There is a private key known as priv and the public key is the point of the graph and is plotted priv amount of times. Since ECC is so efficient and doesn't take much computational power, it is used on a wide variety of weaker computational systems like smart phones. ECC uses a much smaller key size when compared to other asymmetric algorithms, but are much harder to break.

WEAK / DEPRECATED ALGORITHMS

A cryptographic algorithm is usually well known and tested which builds confidence in its ability to protect data. Weak or deprecated is when the algorithm is brute force-able, when a breakthrough happens or any other discovery or test is found that compromises the integrity of encrypted or hashed data.

KEY EXCHANGE

When two parties securely exchanging cryptographic keys over a public channel.

DIGITAL SIGNATURES

Digital signatures are like fingerprints. They are unique identities for a user so you know that if a document, file, or other piece of data is signed it has to be from them.

DIFFUSION VS CONFUSION

These terms are commonly used to describe ciphers and their differences should be known.

DIFFUSION: This means that if a character in plaintext is changed, then several characters in the ciphertext are changed and vice versa.

CONFUSION: This means that the key doesn't relate one to one with the ciphertext. Each character within the ciphertext relies on multiple parts of the key.

STEGANOGRAPHY

Steganography is hiding secret information within something else. It is commonly used to describe hiding messages in photos.

OBFUSCATION

Obfuscation is the act of making data hard to read or understand. It can be done a variety of different ways and is commonly known to be a weak way to hide information.

KEY STRENGTH

Typically based on the amount of bits a key size is and also how difficult it is to generate a key. Keys are typically considered weaker as computational power increases and more bits are needed to prevent the key size from making an algorithm brute forcible.

SESSION & EPHEMERAL KEYS

A Session Key is a cryptographic key that is randomly generated for a given session between a user and a computer. They are also referred to as Symmetric Keys since one is used to encrypt and decrypt.

A key is considered ephemeral if is generated for each time it is needed. Since session keys are generated for each time they are used. They are known as ephemeral.

SECRET ALGORITHM

The idea that algorithms should be kept private to ensure an exploit isn't found to compromise the data they are protecting. Many would argue though that a well-tested algorithm that is well-known with high integrity is more secure than a secret algorithm.

DATA-IN-TRANSIT VS DATA-AT-REST VS DATA-IN-USE

DATA-IN-TRANSIT: The data is moving from one location to another location like from router to router in a public or private network. This is when data is the least secure and when data protection needs to be critical. Data is typically secured by agreeing on the encryption platform to use on both ends of the connection and following the protocols defined by the platform.

DATA-AT-REST: Data that is sitting on a computer and not actively moving. This could be data on a hard drive or archived in some way. Data-at-Rest does not need to be accessed by anybody which makes it less vulnerable than the other two depending on the situation. It is typically secured through perimeter defenses like firewalls and encrypting the storage unit.

DATA-IN-USE: Data that needs to be accessed by somebody and is at a higher risk of vulnerability based on the amount of people that need to view it. Data-in-Use is typically secured through access control policies and authentication.

RANDOM / PSEUDO-RANDOM & NUMBER GENERATION

True randomness is almost never achievable. In computation, random is typically generated using something unpredictable and physically external to a system, like atmospheric noise, while pseudo-random is generated using an algorithm with something like the current system time.

These two methods can be used to generate numbers. They are typically referred to as True Random Number Generators (TRNGs) Pseudo-Random Number Generators (PRNGs).

KEY STRETCHING

This concept makes a cryptographic key stronger by running it through an algorithm.

IMPLEMENTATION VS ALGORITHM SELECTION

The strength and reliability of a cryptosystem is not based solely on the algorithms in use. There is also the concern of the software code used to implement the algorithm. It is important to consider the quality of the code and the reputation of the vendor, as well as the algorithms offered.

A Crypto Service Provider, CSP, is a software library that implements the CAPI. CAPI and CSP serve to provide standardized cryptographic functions to applications.

A crypto module is a hardware or software component that can be used to provide cryptographic services to a device, application, and operation system. A crypto module may provide random number generation, perform hashing, perform encryption and decryption functions, and serve as a secure storage container for encryption keys.

PERFECT FORWARD SECRECY

Perfect Forward Secrecy is commonly used on the transport layer and is a key agreement that makes your session key safe even if the server's private key being used is compromised. A unique session key is also generated for each session a user starts so even losing one key only compromises the data from that specific session.

SECURITY THROUGH OBSCURITY

Security through obscurity is when the design or implementation of something relies on secrecy for security. It is considered a bad idea because nothing is really put in place to stop someone that is dedicated from compromising the system.

COMMON USE CASES

LOW-POWER DEVICES: Smartphones, tablets, and some notebook computers, may have limited CPU capabilities or memory capacities. These devices require cryptography functions that will not place an undue burden of computation on the device and will also minimize latency and delay caused by heavy computational loads.

LOW-LATENCY SYSTEMS: Those that require real-time or near-real-time response and communications, such as navigation, VoIP, and some high-end web services. Encrypting and decrypting communications takes time and effort, so in order to minimize latency many devices have a dedicated crypto-processor that will offload the cryptographic operations from the CPU in order to provide faster security services.

HIGH-RESILIENCY SYSTEMS: These want to ensure reliable communications and data storage, often at the expense of higher latency and by requiring more computational capabilities, such as banking and military weapons control systems.

CONFIDENTIALITY: Protects the secrecy of data, information, or resources. It prevents or minimizes unauthorized access to data. It ensures that no one other than the intended recipient of a message receives it or is able to read it.

INTEGRITY: The security service that protects the reliability and correctness of data. Integrity protection prevents unauthorized alterations of data. It ensures that data remains correct, unaltered, and preserved.

OBFUSCATION: There may be some circumstances or situations where obfuscation is more desirable than actual encryption. These might include communications that are valid or relevant only for a second or two so that the time and effort to perform encryption is greater than the risk of disclosure during the small time frame of value.

AUTHENTICATION: This is often a key element of a cryptography solution. When sending secured communications, it is essential to control who the recipient is and to verify the recipient before disclosing the content.

NONREPUDIATION: This prevents the senders of a message or the perpetrators of an activity from being able to deny that they sent the message or performed the activity.

> **NOTE**: Cryptography itself may be constrained by resource limitations or security policy. If resources are limited or constrained, then cryptography may be restricted or unable to operate. A company security policy or government legislation may prohibit certain uses of cryptography, such as prohibiting the use of a VPN or requiring that all communications be content filtered by an IDS/IPS and firewall.

CRYPTOGRAPHY ALGORITHMS & THEIR BASIC CHARACTERISTICS

SYMMETRIC ALGORITHMS

Symmetric Algorithms are those that use one key to both encrypt and decrypt. Examples of these algorithms are outlined below:

ADVANCED ENCRYPTION STANDARD: AES uses key lengths of 128, 192, and 256 bits and block sizes of 128 bits. It uses substitution and permutation on data in matrix form and is a fast algorithm in both software and hardware. As of 2018, AES is widely used in the US government and worldwide and surpassed its predecessor, DES, due to having a higher key size and better algorithm. It was originally named Rijndael, but was officially renamed in 2000 after the NIST accepted it as the new Advanced Encryption Standard.

DATA ENCRYPTION STANDARD: DES is now deemed insecure, but was an integral part of cryptography. It uses 56-bit keys and a block size of 64 bits. DES takes plaintext and runs it through a block cipher: a series of complex equations constructed in a feistel network.

3DES: A block cipher that runs the DES cipher on a data block three times.

RIVEST CIPHER 4: RC4 generates a pseudorandom set of bits known as a keystream and combines it with plaintext using a bitwise operator.

BLOWFISH: A block cipher developed in 1993 that was used in many applications and is known to be fast. (AES and Twofish, as of 2018, are the current algorithms used in its place). Blowfish has key sizes between 32 and 448 bits and a block size of 64 bits. It uses the feistel network like DES and does 16 rounds.

TWOFISH: Has a block size of 128 bits and key sizes of up to 256 bits. It was in the finals to be the Advanced Encryption Standard, but didn't quite make the cut. It also uses a feistel network and does 16 rounds. It is well known because of the complexity of its key schedule.

CIPHER MODES

Cipher Modes, also known as mode of operations, are algorithms that use a block cipher to provide secrecy or authenticity. Stream algorithms encrypt data byte by byte while block algorithms encrypt data block by block (where a block is multiple bytes defined by the program). Both are used in cipher modes and some Cipher Modes are outlined below:

CIPHER BLOCK CHAINING: CBC is not encryption parallelizable, but is decryption parallelizable and has random read access. It is the most commonly used mode of operation. It takes in plain text, splits it into blocks, then XORs with an IV and uses a key to block cipher to make ciphertext. Each ciphertext for each block is XOR'd with the previous before encryption, creating a chain like effect that dubbed its name. Its main drawback is that it's sequential and needs padding.

ELECTRONIC COOKBOOK: ECB is encryption parallelizable, decryption parallelizable, and has random read access. It is the simplest used mode of operation. Electronic Cookbook breaks a message into blocks, takes in plaintext, then uses a key with a block cipher to make ciphertext. Each plaintext block is converted into ciphertext blocks, so there's no diffusion as other methods. Data patterns are not hidden and it is not recommended for any real cryptographic protocol.

COUNTER: CTR turns block ciphers into stream ciphers. It is encryption parallelizable, decryption parallelizable, and has random read access. It splits the data into blocks and then uses a series of nonces with a counter and feeds both into a block cipher with a key. Then it takes the plaintext and XORs it with the output from the block cipher to make the ciphertext.

GALOIS / COUNTER MODE: GCM has been adopted for its speed and performance. GCM functions like a CTR mode, but combines the ciphertext at the end with an authentication code through XOR so that it has an auth tag to verify the data.

ASYMMETRIC ALGORITHMS

RSA: A public key encryption scheme that creates a signature that relies on the computational difficulty of factoring large numbers to find the two original prime factors. Alice creates a hash digest of her payload, encrypts it with her RSA private key and then sends it to Bob. Bob would then decrypt the payload with Alice's public key to verify that it came from Alice.

DIGITAL SIGNATURE ALGORITHM: DSA was created for digital signatures. DSA's security is based on discrete logarithms and is used for signing payloads. DSA offers forward secrecy.

THE DIFFIE-HELLMAN ALGORITHM: This relied on Alice and Bob creating a shared private key. The private key is used as an encryption key for encrypting and sending payloads to one another on an open communications channel. Groups determine the strength of the key used in the key exchange process. The higher the group number, the more secure, but the more time required for key computation. Peers within the same session are required to use the same group.

DIFFIE-HELLMAN EPHEMERAL: A flavor of Diffie-Hellman where every conversation has a new and temporary key generated. This way the same key is never generated twice and provides forward secrecy, thus all past conversations are safe if a single key is leaked.

ELLIPTIC-CURVE DIFFIE-HELLMAN: ECDHE is a Diffie-Hellman Exchange where the exchange is signed with an RSA signature. ECDHE offers faster performance and smaller messages than normal DHE as a 224-bit curve offers the same security as 2048-bit plain DH Key.

ELLIPTIC CURVE CRYPTOGRAPHY: Public key cryptography based on the algebraic structure of elliptic curves, usually given by the equation $y^2 = x^3 + ax + b$ over finite fields. Elliptic curves are used for key agreements, digital signatures, pseudo-random generators and other tasks. This can be seen in Elliptic Curve Diffie-Hellman and Elliptic Curve Digital Signature Algorithm.

PRETTY GOOD PRIVACY: PGP is an encryption method for emails and files. PGP encryption uses a combination of encryption methodologies such as hashing, data compression, symmetric-key cryptography and public key cryptography to keep data secure. PGP has been formalized as the OpenPGP standard.

GNU PRIVACY GUARD: GPG is a rewrite of PGP using AES encryption rather than IDEA encryption. The reason for the change is AES's higher performance and more security when compared to the IDEA encryption algorithm used in PGP. GPG is not patented and is available royalty free.

HASHING ALGORITHMS

MESSAGE-DIGEST-5: MD5 is a one-way cryptographic hash function where a message of any length is input and creates a 128-bit output of a fixed-length digest value that is used to authenticate the original message. MD5 is intended to be used for digital signature applications, where a large file must be 'compressed' in a secure manner before being encrypted with a private key under a public-key cryptosystem such as RSA.

SECURE HASH ALGORITHMS: SHA is a family of cryptographic hash functions published by the National Institute of Standards and Technology as a U.S. Federal Information Processing Standard, or FIPS.

SHA-1: A 160-bit hash function that resembles MD5. It was designed by the National Security Agency. It is no longer in use due to a large security flaw discovered in 2010. SHA-2 is a family of hashing algorithms with different block sizes, SHA-256 and SHA-512. SHA-256 uses 32-bit words and SHA-256 uses 64 bits. SHA-3 is the most recent member of SHA and is derived from the Keccak hash algorithm. It supports the same hash lengths as SHA-2, and its internal structure differs significantly from the rest of the SHA family.

KEYED-HASHING FOR MESSAGE AUTHENTICATION: HMAC is a message authentication code obtained by running a hash function like MD5, SHA1, etc. over the data and a shared private key. When two parties exchange messages through those secure file transfer protocols, those messages will be accompanied by HMACs instead of plain hashes. An HMAC employs both a hash function and a shared secret key.

SHARED SECRET KEY: Provides exchanging parties a way to establish message authenticity. It provides the two parties a way of verifying both the message and HMAC for authentication.

RACE INTEGRITY PRIMITIVES EVALUATION MESSAGE DIGEST: RIPEMD is a family of cryptographic hash functions. RIPEMD-160 is used in the Bitcoin standard. It is a strengthened version of the RIPEMD algorithm which produces a 128-bit hash digest while the RIPEMD-160 algorithm produces a 160-bit output. The compression function is made up of 80 stages made up of 5 blocks that run 16 times each. This pattern runs twice with the results being combined at the bottom using modulo 32 addition.

KEY STRETCHING ALGORITHMS

BCRYPT: A password hashing function designed by Niels Provos and David Mazières in 1999. BCRYPT wants the hash function to be slow for attackers, but not the user. This algorithm can modify the key factor, making the hash N times more difficult to hash, slowing down the attacker.

PASSWORD-BASED KEY DERIVATION FUNCTION 2: PBKDF2 is a key derivation functions with a sliding computational cost aimed to reduce the vulnerability of encrypted keys to brute force attacks. PBKDF2 applies a pseudorandom function to the input along with a salt and repeats the process a number of times to generate a derived key. The derived key is used as a cryptographic key in subsequent operations.

OBFUSCATION

XOR CIPHER: A type of additive cypher and follows a set of rules when XORing an input. It is very common in complex ciphers. A XOR cipher is easy to break and inexpensive computationally. The only benefit of XOR is how easy it is to implement.

XOR follows the following rules:
- A XOR 0 = A
- A XOR A = 0
- (A XOR B) XOR C = A XOR (B XOR C)
- (B XOR A) XOR A = B XOR 0 = B

ROT13: A substitution cipher that replaces a letter with the 13th letter after it in the alphabet. The same method is used to decrypt ROT ciphers. It is not computationally difficult to solve and is known to be weak encryption.

SUBSTITUTION CIPHER: An encryption scheme where units of the plaintext are replaced by other symbols or groups of symbols. For every instance of 'A' replace it with 'D', so on and so forth. The Caesar Cipher, ROT13, and Vigenere Cipher are examples.

GIVEN A SCENARIO, INSTALL & CONFIGURE WIRELESS SECURITY SETTINGS

CRYPTOGRAPHIC PROTOCOLS

WIFI PROTECTED ACCESS: WPA was adopted in 2003 by the WiFi Alliance due to significant vulnerabilities in WEP. WEP uses integrity and TKIP checks to ensure data has not been tampered with when transferred between the client and provider.

TEMPORAL KEY INTEGRITY PROTOCOL: TKIP is an 8 to 63 character long passphrase paired with a network SSID to generate a unique encryption key for each wireless client. These keys are constantly changed, but static keys are able to be generated. The most common configuration is WPA-PSK, WPA Pre-Shared Key. WPA-PSK is designed for home and small networks that do not rely on an authentication server.

WIFI PROTECTED ACCESS 2: WPA2 provides both stronger data protection and network access control than WPA. WPA2 uses FIPS 140-2 compliant AES encryption algorithm and 802.1x-based authentication. Two versions of WPA2 exist: WPA2-Personal and WPA2-Enterprise. WPA2-Personal protects unauthorized network access by utilizing a set-up password. WPA2-Enterprise verifies network users through a server. WPA2 is backward compatible with WPA. Known WPA2 vulnerabilities are limited almost entirely to enterprise networks and have little to no practical consideration in regard to home network security.

COUNTER MODE WITH CIPHER BLOCK CHAINING MESSAGE AUTHENTICATION CODE PROTOCOL: CCMP is an encryption protocol designed for Wireless LAN products. CCMP is the standard encryption protocol for WPA2 and uses CTR for confidentiality and CBC-MAC for authentication and integrity. It will most often be seen when you're setting up your wireless network as WPA2. CCMP uses CTR for data confidentiality, where only authorized users are allowed to receive information across a network. There's also authentication enabled within CCMP using CBC-MAC; users can be assured that the other users on the network really are genuine users. There's also access control implemented within CCMP, where users are able to allow or disallow access to the network based on your credentials.

TKIP was endorsed by the WiFi Alliance on October 31, 2002 under the name WiFi Protected Access, WPA. TKIP uses a key mixing function that combines the secret root with the initialization vector, a pseudorandom fixed-size input for the cryptographic primitive, before passing it to the RC4 cipher initialization. This provides protection against related-key attacks. TKIP includes sequence counters to protect against replay attacks. In a replay attack, the attacker records the information being sent over a network and will replay that information in an attempt to gain access. Since old information has old counters, it is immediately rejected by TKIP.

When configuring WPA encryption on your router first go to your security settings. You will see the following options Security Mode, WPA Mode, Cipher Type, Group Key Update Interval and Pre-Shared Key. AES is the cipher type.

The Preshared Key (PSK) is used to enable connectivity between wireless devices. PSK is automatically used when you select WPA-Personal in the Security Mode section. You may want to write this down as you'll need to enter the same key when configuring your network card.

The other option is WPA-Enterprise, which uses a RADIUS server in this AP. So, if you ever see the term "WPA2-PSK," that means that the AP is set up to use the WPA2 protocol with a preshared key, and not an external authentication method such as RADIUS.

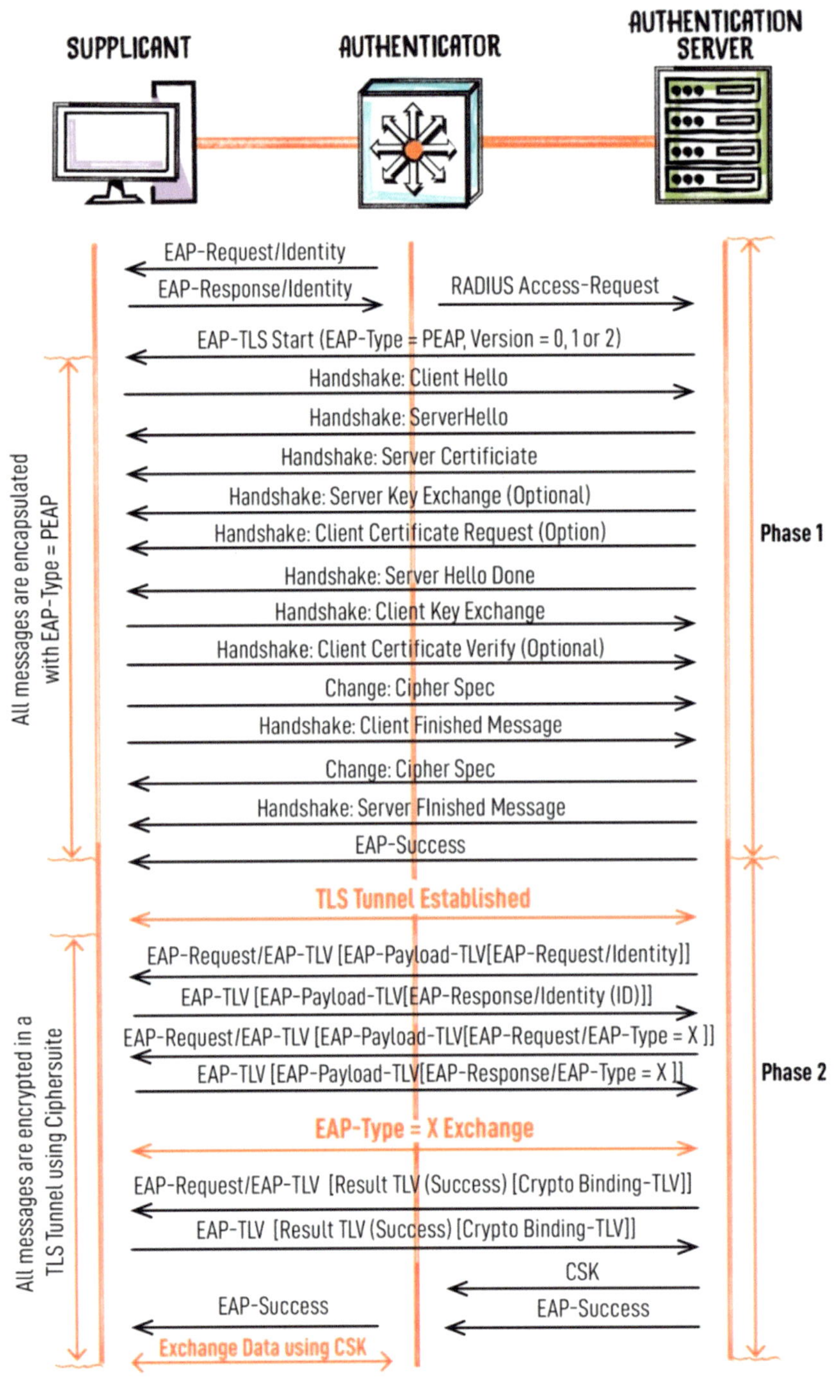
SUPPLICANT
AUTHENTICATOR
AUTHENTICATION SERVER
EAP-Request/Identity
EAP-Response/Identity
RADIUS Access-Request
EAP-TLS Start (EAP-Type = PEAP, Version = 0, 1 or 2)
Handshake: Client Hello
Handshake: ServerHello
Handshake: Server Certificiate
Handshake: Server Key Exchange (Optional)
Handshake: Client Certificate Request (Option)
Handshake: Server Hello Done
Handshake: Client Key Exchange
Handshake: Client Certificate Verify (Optional)
Change: Cipher Spec
Handshake: Client Finished Message
Change: Cipher Spec
Handshake: Server FInished Message
EAP-Success
All messages are encapsulated with EAP-Type = PEAP
Phase 1
TLS Tunnel Established
EAP-Request/EAP-TLV [EAP-Payload-TLV[EAP-Request/Identity]]
EAP-TLV [EAP-Payload-TLV[EAP-Response/Identity (ID)]]
EAP-Request/EAP-TLV [EAP-Payload-TLV[EAP-Request/EAP-Type = X]]
EAP-TLV [EAP-Payload-TLV[EAP-Response/EAP-Type = X]]
EAP-Type = X Exchange
EAP-Request/EAP-TLV [Result TLV (Success) [Crypto Binding-TLV]]
EAP-TLV [Result TLV (Success) [Crypto Binding-TLV]]
CSK
EAP-Success
EAP-Success
All messages are encrypted in a TLS Tunnel using Ciphersuite
Phase 2
Exchange Data using CSK

EXTENSIBLE AUTHENTICATION PROTOCOL: EAP is a frequently used authentication framework in wireless networks and point-to-point connections. EAP is often used when a computer is connecting to the internet. It supports multiple authentication methods such as one-time passwords, token cards, certificates, public key authentication and smart cards. In wireless communications the user is requesting a connection to a WLAN through the access point. Upon request the AP requests the identity of the user to an authentication server. The server asks the AP for proof of identity and once the exchange of identifications is complete, the user is authorized or rejected.

PROTECTED EXTENSIBLE AUTHENTICATION PROTOCOL: PEAP corrects deficiencies in traditional EAP. EAP assumes that the channel where the exchange occurs is secure. PEAP ensures a secure channel by encapsulating EAP within a potentially encrypted and authenticated TLS tunnel. This ensures protection for the EAP conversation.

EAP-FAST: A protocol proposed by Cisco to fix the weaknesses in LEAP without losing the lightweight of it. This uses a protected access credential instead of a certificate to achieve mutual authentication. FAST stands for flexible authentication via secure tunneling.

EAP-FAST has 3 phases. Phase 0 uses Authenticated Diffie-Hellman Protocol to provide a peer with a shared secret to eliminate the requirement of the peer establishing a master secret key every time they require network access. Phase 1 establishes a tunnel for the conversation using a tunnel key. This key establishment provides confidentiality and integrity during the authentication process. In phase 2 the server authenticates via multiple tunneled, authentication mechanisms.

EAP TRANSPORT LAYER SECURITY: Provides for certificate-based and mutual authentication of the client and the network. It relies on client-side and server-side certificates to perform authentication and can be used to dynamically generate user-based and session-based WEP keys to secure subsequent communications between the WLAN client and the access point. One drawback of EAP-TLS is that certificates must be managed on both the client and server side. For a large WLAN installation, this could be a very cumbersome task.

TUNNELED TRANSPORT LAYER SECURITY: EAP-TTLS was developed as an extension of EAP-TLS. This security method provides for certificate-based, mutual authentication of the client and network through an encrypted tunnel, as well as a means to derive dynamic, per-user, per-session WEP keys.

: EAPOL, or IEEE 802.1x, was developed to give network sign-on capabilities to access network resources. It's a simple encapsulation of EAP over a wired connection that still requires a supplicant, authenticator, and authentication server for an authenticated conversation.

It all starts with the central connecting device, such as a switch or wireless access point. These devices must first enable 802.1X connections; they must have the 802.1X protocol (and supporting protocols) installed. Vendors that offer 802.1x compliant devices (for example, switches and wireless access points) include Cisco, Symbol Technologies, and Intel. Next, the client computer needs to have an operating system, or additional software, that supports 802.1X. The client computer is known as the supplicant. All recent Windows versions support 802.1X, including Windows 10, 8, 7, Vista, and XP, though each comes with its own advantages and disadvantages. OS X offers support as well, and Linux computers can use Open1X to enable client access to networks that require 802.1X authentication.

RADIUS FEDERATION: When an organization has multiple RADIUS servers, possibly on different networks, which need to communicate with each other in a safe way. It is accomplished by creating trust relationships and developing a core to manage those relationships as well as the routing of authentication requests. It is often implemented in conjunction with IEEE 802.1x. This federated network authentication could also span between multiple organizations. Remember, RADIUS uses ports 1812 and 1813 (or 1645 and 1646).

METHODS: PSK VS ENTERPRISE VS OPEN

OPEN WIRELESS NETWORK: Not recommended. It has no authentication and has a lack of encryption. Most public wireless network scenarios use open networks to minimize hassle and troubleshooting for users. If you have to use an open network, try using a VPN to encrypt all of your personal traffic from your device.

PRE SHARED KEY: PSK is where two parties share a key via an out-of-band communication method prior to actual communication. This is where PSK over WEP becomes a problem, because the same value used for encryption is also used for authentication for all users. PSK over WPA or WPA2 still uses a fixed value for all users, but a password/phrase can be used to a stronger effect. PSK should be used when Enterprise is not available.

ENTERPRISE: ENT enables the leveraging of an existing AAA Service, such as RADIUS, to be used for authentication. This enables unique authentication per connected user / device rather than a fixed shared authentication like with PSK. ENT should always be used wherever leveraging a AAA service is available.

WIFI PROTECTED SETUP: WPS is in of itself a security vulnerability. It was originally created to provide users with easy connectivity to a wireless access point and was later suggested by all major manufacturers that it be disabled. The problem with WPS was the eight-digit code. It effectively worked as two separate smaller codes that collectively could be broken by a brute-force attack within hours. There isn't much that can be done to prevent the problem other than disabling WPS altogether in the access point's firmware interface or, if WPS can't be disabled, upgrading to a newer device. WPS is a deprecated and insecure technology that should not be allowed on a wireless network.

CAPTIVE PORTALS: Have you ever had to login to a Starbucks, Target or a hotel? Usually you will be redirected to a page requiring authentication before being able to use their internet. You may have to create an account or enter a password provided by the business. A captive portal forces the web browser of the device, laptop, or phone to authenticate itself. The redirect may happen through HTTP, HTTPS, or DNS services. More often than not, it'll be through HTTPS.

How does an attacker circumvent this and how do we protect against it? To protect the network, a company can offer a multifactor authentication method. There are many free applications the company can use for Windows and Linux platforms. The whole point of the technology is the ability to track users who access the free wireless network. If the user performs any suspect actions, they can be traced by way of email address, IP address, and MAC address, in addition to other means if multifactor authentication is used.

GIVEN A SCENARIO, IMPLEMENT PUBLIC KEY INFRASTRUCTURE

COMPONENTS

A CERTIFICATE AUTHORITY: CA is a trusted entity that issues electronic documents that verify a digital entity's identity on the Internet (i.e. Global Sign).

INTERMEDIATE CA: CA that is subordinate to the root CA by one or more levels and typically issues certificates to other CAs in the public key infrastructure (PKI) hierarchy.

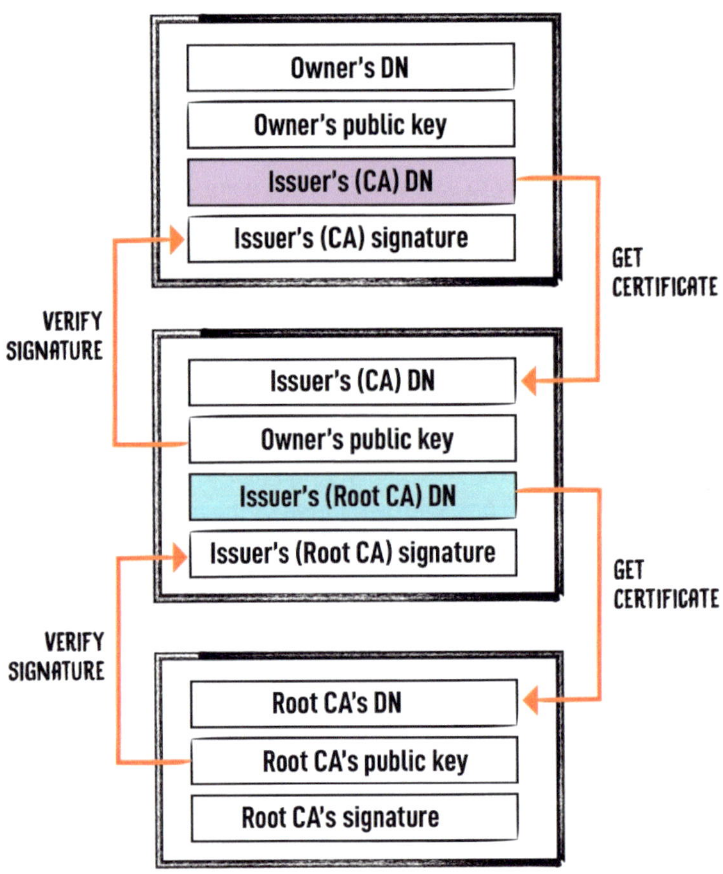

CERTIFICATE REVOCATION LIST: CRL is a blacklist of digital certificates that have been revoked by the issuing Certificate Authority before their scheduled expiration date and should no longer be trusted. This is usually done through an Anti-Certificate.

OCSP: An Internet protocol used for obtaining the revocation status of an X.509 digital certificate. It was developed to be an alternative to a CRL.

CERTIFICATE SIGNING REQUEST: CSR is a message sent from an applicant to a certificate authority in order to apply for a digital identity certificate.

A PUBLIC KEY CERTIFICATE: Also called a digital certificate, this electronic document is used to prove the ownership of a public key usually issued by a Certificate Authority.

PUBLIC KEYS: Keys available for everyone to see. Public keys are used to convert a message into an unreadable format.

PRIVATE KEY: Also known as a secret key, this is a variable that is used with an algorithm to encrypt and decrypt code. Quality encryption always follows a fundamental rule: the algorithm doesn't need to be kept secret, but the key does.

OBJECT IDENTIFIERS: OID are numeric values that enable programs to determine whether a certificate is valid for a particular use. The object identifiers can be used to represent components such as X509 extensions, PKCS #7 extensions, and PKCS #7 contents. Each subtree in the Microsoft object identifier is assigned to a specific area.

CONCEPTS: ONLINE VS OFFLINE CA

OFFLINE CA: This is a root CA of a hierarchy that is disconnected from the network and often powered off to be stored in a vault, safe, or other physically secured container. The Offline CA is only brought online to issue more certificates to intermediary, subordinate leaf CAs in the case of expiration or near expiration.

ONLINE CA: These are always online and network connected. The purpose of keeping a CA offline is to keep the CA from being compromised because if the Root CA is compromised, then the entire hierarchy is compromised. They are kept secure physically and logically in order to manage the risk associated with a high availability Online CA.

So which is more secure? We would immediately think the Offline CA is more secure because if it is not available to be attacked, then how can an attacker compromise it? Over the years, this technique has become not as reliable as one would think. DigiNotar, a Dutch company, had an Offline CA and on July 10, 2011, an attacker with access to DigiNotar's systems issued a wildcard certificate for Google. This certificate was subsequently used by unknown persons in Iran to conduct a man-in-the-middle attack against Google services. A recommended alternative is to use a hardware security module to protect the authentication services and encryption keys of all CA systems.

OCSP STAPLING: A means of checking the revocation status of X.509 digital certificates. This mechanism enables the presenter of a certificate to append or staple a time stamp OCSP response signing by issuing CA. This stapling process allows the client or recipient to verify the revocation status of the offered digital certificate without having to interact with the issuing CA's OCSP solution. It also reduces the workload for the client and server and minimizes the risk of a DoS against the OCSP system. This could force the client into a default-accept mode if no OCSP response was received.

HTTP PUBLIC KEY PINNING: Pinning is a security mechanism operating over HTTP that enables an HTTPS system to prevent impersonation by attackers through fraudulently issued digital certificates. Pinning operates by providing the visitor with an HTTP response header field value labeled Public Key. Public Key-Pins use hashes of the certificates used by the server along with a time stamp for how long to keep these certificates pinned. If the connection is secure, the client will receive a list of valid certificates to accept from or for the specific server. If the client revisits the same server, the server will compare the offered certificate hashes and the pinned certificate hashes. A match will result in interaction with the website, otherwise, the connection is rejected.

TYPES OF CERTIFICATES

WILDCARD CERTIFICATE: A public key certificate that can be used on unlimited subdomains of a domain i.e. "*.example.com". The principal use is for securing web sites with HTTPS.

SAN CERTIFICATE: Allows for multiple domains to be protected. If you have a cert for "example.com", the same cert can be used for "example.net" and "example.org". SAN values can be changed at any time to fit the situation.

CODE SIGNING: A method of using certificates to digitally sign executables and scripts in order to verify the author's identity and ensure the code hasn't been altered since its signing. Code Signing provides authentication and Integrity.

SELF-SIGNED CERTIFICATE: An identity certificate signed by the entity whose identity it certifies. Unlike Certification Authority issued certificates, these are free of charge.

MACHINE / COMPUTER CERTIFICATE: Issued to verify the identity of the device used rather than a service or a user.

EMAIL CERTIFICATE: Used to verify an email address.

USER CERTIFICATE: Used to verify a specific individual person.

ROOT: A public key certificate that identifies a root certificate authority (CA). A CA can issue multiple certificates from a tree structure. The root certificate is usually made trustworthy by some mechanism other than a certificate, such as by secure physical distribution.

DOMAIN VALIDATION: An encryption-only certificate, Domain Validation Certificate is used for TLS. It certifies that the owner of the domain has control over the domain. This certification does not assure that any legal entity is issuing the certificate.

EXTENDED VALIDATION CERTIFICATE: Used for HTTPS websites and software. This certificate proves that a legal entity controls the website or software package. Obtaining an Extended Validation requires verification of the requesting entity by a certificate authority.

CERTIFICATE FORMATS

DER EXTENSION: Used for binary DER encoded certificates.

PEM EXTENSION: Used for different types of X.509v3 files which contain ASCII (Base64) armored data prefixed with a "–– BEGIN ..." line.

PFX EXTENSION: Used for storing the Server certificate, any intermediate certificates & private key in one encryptable file.

CER EXTENSION: For an SSL certificate file format, this is used by Web servers to help verify the identity and security of the site in question. SSL certificates are provided by a third-party security certificate authority such as VeriSign, GlobalSign or Thawte.

P12 EXTENSION: P12 is the combined format that holds the private key and certificate and is the format most modern signing utilities use. Also referred to as a PFX file.

P7B: This contains "––BEGIN PKCS7––" & "––END PKCS7––" statements. It can contain only certificates & chain certificates but not the private key. They are Base64 encoded ASCII files.

NOTES

ACRONYMS SPELLED OUT

3DES	Triple Digital Encryption Standard
AAA	Authentication, Authorization, and Accounting
ACL	Access Control List
AES	Advanced Encryption Standard
AES256	Advanced Encryption Standards 256bit
AH	Authentication Header
ALE	Annualized Loss Expectancy
AP	Access Point
API	Application Programming Interface
ASP	Application Service Provider
ARO	Annualized Rate of Occurrence
ARP	Address Resolution Protocol
AUP	Acceptable Use Policy
BCP	Business Continuity Planning
BIOS	Basic Input / Output System
BOTS	Network Robots
CA	Certificate Authority
CAC	Common Access Card
CAN	Controller Area Network
CCMP	Counter-Mode/CBC-Mac Protocol
CCTV	Closed-circuit television
CERT	Computer Emergency Response Team
CHAP	Challenge Handshake Authentication Protocol
CIRT	Computer Incident Response Team

CMM	Capability Maturity Model
COOP	Continuity of Operation Planning
CP	Contingency Planning
CRC	Cyclical Redundancy Check
CRL	Certification Revocation List
CSU	Channel Service Unit
DAC	Discretionary Access Control
DDOS	Distributed Denial of Service
DEP	Data Execution Prevention
DES	Digital Encryption Standard
DHCP	Dynamic Host Configuration Protocol
DLL	Dynamic Link Library
DLP	Data Loss Prevention
DMZ	Demilitarized Zone
DNS	Domain Name Service (Server)
DOS	Denial of Service
DRP	Disaster Recovery Plan
DSA	Digital Signature Algorithm
DSL	Digital Subscriber line
DSU	Data Service Unit
EAP	Extensible Authentication Protocol
ECC	Elliptic Curve Cryptography
EFS	Encrypted File System
EMI	Electromagnetic Interference

ESP	Encapsulated Security Payload
FTP	File Transfer Protocol
GPO	Group Policy Object
GPU	Graphic Processing Unit
GRE	Generic Routing Encapsulation
HDD	Hard Disk Drive
HIDS	Host Based Intrusion Detection System
HIPS	Host Based Intrusion Prevention System
HMAC	Hashed Message Authentication Code
HSM	Hardware Security Module
HTML	Hypertext Markup Language
HTTP	Hypertext Transfer Protocol
HTTPS	Hypertext Transfer Protocol over SSL
HVAC	Heating, Ventilation Air Conditioning
IaaS	Infrastructure as a Service
ICMP	Internet Control Message Protocol
ID	Identification
IKE	Internet Key Exchange
IM	Instant messaging
IMAP4	Internet Message Access Protocol v4
IP	Internet Protocol
IPSEC	Internet Protocol Security
IRC	Internet Relay Chat
ISP	Internet Service Provider

ITCP	IT Contingency Plan
IV	Initialization Vector
KDC	Key Distribution Center
L2TP	Layer 2 Tunneling Protocol
LAN	Local Area Network
LANMAN	Local Area Network Manager
LDAP	Lightweight Directory Access Protocol
LEAP	Lightweight Extensible Authentication Protocol
MAC	Mandatory Access Control / Media Access Control
MAC	Message Authentication Code
MAN	Metropolitan Area Network
MBR	Master Boot Record
MD5	Message Digest 5
MPLS	Multi-Protocol Layer Switch
MSCHAP	Microsoft Challenge Handshake Authentication Protocol
MTBF	Mean Time Between Failures
MTTR	Mean Time to Recover
MTU	Maximum Transmission Unit
NAC	Network Access Control
NAT	Network Address Translation
NDA	Non-Disclosure Agreement
NIDS	Network Based Intrusion Detection System
NIPS	Network Based Intrusion Prevention System
NIST	National Institute of Standards & Technology

NOS	Network Operating System
NTFS	New Technology File System
NTLM	New Technology LANMAN
NTP	Network Time Protocol
OCSP	Online Certification Security Protocol
OLA	Open License Agreement
OS	Operating System
OVAL	Open Vulnerability Assessment Language
PAM	Pluggable Authentication Modules
PAP	Password Authentication Protocol
PAT	Port Address Translation
PBX	Private Branch Exchange
PCAP	Packet Capture
PEAP	Protected Extensible Authentication Protocol
PED	Personal Electronic Device
PGP	Pretty Good Privacy
PII	Personally Identifiable Information
PIV	Personal Identity Verification
PKI	Public Key Infrastructure
POTS	Plain Old Telephone Service
PPP	Point-to-point Protocol
PPTP	Point to Point Tunneling Protocol
PSK	Pre-Shared Key
PTZ	Pan-Tilt-Zoom

RA	Recovery Agent
RAD	Rapid application development
RADIUS	Remote Authentication Dial-in User Server
RAID	Redundant Array of Inexpensive Disks
RAS	Remote Access Server
RBAC	Role Based Access Control
RBAC	Rule Based Access Control
RIPEMD	RACE Integrity Primitives Evaluation Message Digest
ROI	Return of Investment
RPO	Recovery Point Objective
RSA	Rivest, Shamir, & Adleman
RTO	Recovery Time Objective
RTP	Real-Time Transport Protocol
S/MIME	Secure / Multipurpose internet Mail Extensions
SAML	Security Assertions Markup Language
SaaS	Software as a Service
SCAP	Security Content Automation Protocol
SCSI	Small Computer System Interface
SDLC	Software Development Life Cycle
SDLM	Software Development Life Cycle Methodology
SEH	Structured Exception Handler
SHA	Secure Hashing Algorithm
SHTTP	Secure Hypertext Transfer Protocol
SIM	Subscriber Identity Module

SLA	Service Level Agreement
SLE	Single Loss Expectancy
SMS	Short Message Service
SMTP	Simple Mail Transfer Protocol
SNMP	Simple Network Management Protocol
SOAP	Simple Object Access Point
SONET	Synchronous Optical Network Technologies
SPIM	Spam over Internet Messaging
SSD	Solid State Drive
SSH	Secure Shell
SSL	Secure Sockets Layer
SSO	Single Sign On
STP	Shielded Twisted Pair
TACACS	Terminal Access Controller Access Control System
TCP/IP	Transmission Control Protocol / Internet Protocol
TKIP	Temporal Key Integrity Protocol
TLS	Transport Layer Security
TPM	Trusted Platform Module
TSIG	Transaction Signature
UAT	User Acceptance Testing
UEFI	Unified Extensible Firmware Interface
UPS	Uninterruptable Power Supply
URL	Universal Resource Locator
USB	Universal Serial Bus

UTP	Unshielded Twisted Pair
VDI	Virtualization Desktop Infrastructure
VLAN	Virtual Local Area Network
VoIP	Voice over IP
VPN	Virtual Private Network
VTC	Video Teleconferencing
WAF	Web-Application Firewall
WAP	Wireless Access Point
WEP	Wired Equivalent Privacy
WIDS	Wireless Intrusion Detection System
WIPS	Wireless Intrusion Prevention System
WPA	Wireless Protected Access
WTLS	Wireless TLS
XML	Extensible Markup Language
XSRF	Cross-Site Request Forgery
XSRF	Cross-Site Request Forgery
XSS	Cross-Site Scripting

Made in the USA
Monee, IL
07 July 2026

56551650R00121